Y0-EIB-885

Energy, Economics, and the Environment

Conflicting Views of an Essential Interrelationship

AAAS Selected Symposia Series

Published by Westview Press, Inc.
5500 Central Avenue, Boulder, Colorado

for the

American Association for the Advancement of Science
1776 Massachusetts Avenue, N.W., Washington, D.C.

Energy, Economics, and the Environment

Conflicting Views of an Essential Interrelationship

Edited by Herman E. Daly and Alvaro F. Umaña

AAAS Selected Symposium **64**

AAAS Selected Symposia Series

This book is based on a symposium which was held at the 1980 AAAS National Annual Meeting in San Francisco, California, January 3-8. The symposium was sponsored by AAAS Sections K (Social and Economic Sciences) and X (General).

Published in 1981 in the United States of America by
Westview Press, Inc.
5500 Central Avenue
Boulder, Colorado 80301
Frederick A. Praeger, Publisher

Library of Congress Cataloging in Publication Data
Main entry under title:
Energy, economics, and the environment.
(AAAS selected symposium ; 64)
"Based on a symposium . . . held at the 1980 AAAS national annual meeting in San Francisco, California, January 3-8 . . . sponsored by AAAS Sections K (Social and Economic Sciences) and X (General)"--T.p. verso.
Includes bibliographical references.
1. Power resources--Congresses. 2. Economics--Congresses. 3. Environmental policy--Congresses. 4. Energy policy--Congresses. I. Daly, Herman E. II. Umaña, Alvaro F. III. American Association for the Advancement of Science. Section K--Social and Economic Sciences. IV. American Association for the Advancement of Science. Section X--General. V. Series.
TJ163.2.E474 333.79 81-11641
ISBN 0-86531-282-6 AACR2

Printed and bound in the United States of America

About the Book

Economies must adjust to the physical constraints of our planet, and economists must join with other scholars to develop novel ways of evaluating these constraints and the services provided by the natural environment. To this end, Alvaro Umaña gives an overview of the problem of providing economics with a biophysical foundation; Nicholas Georgescu-Roegen explains the importance of matter, as well as energy, in economic valuation and extends his integration of thermodynamics with economics; Bruce Hannon offers a methodology for calculating the energy cost of producing energy; Kenneth Arrow reflects on the ideas of Georgescu-Roegen and Hannon from the standpoint of conventional economics; Robert Costanza shows how to count the energy cost of labor and government services and concludes that, when this is done, the variation in energy intensity per dollar's worth of output is significantly less than previously estimated; and finally, Garrett Hardin faces up to the difficult issue of meeting scarcity by controlling demand rather than supply. In a postscript, Herman Daly reviews the main issues and the unresolved problems that gave rise to this volume.

About the Series

The *AAAS Selected Symposia Series* was begun in 1977 to provide a means for more permanently recording and more widely disseminating some of the valuable material which is discussed at the AAAS Annual National Meetings. The volumes in this *Series* are based on symposia held at the Meetings which address topics of current and continuing significance, both within and among the sciences, and in the areas in which science and technology impact on public policy. The *Series* format is designed to provide for rapid dissemination of information, so the papers are not typeset but are reproduced directly from the camera-copy submitted by the authors. The papers are organized and edited by the symposium arrangers who then become the editors of the various volumes. Most papers published in this *Series* are original contributions which have not been previously published, although in some cases additional papers from other sources have been added by an editor to provide a more comprehensive view of a particular topic. Symposia may be reports of new research or reviews of established work, particularly work of an interdisciplinary nature, since the AAAS Annual Meetings typically embrace the full range of the sciences and their societal implications.

WILLIAM D. CAREY
Executive Officer
American Association for the Advancement of Science

Contents

About the Editors and Authors

Herman E. Daly *is a professor of economics at Louisiana State University-Baton Rouge. A specialist in economic development and environmental and resource economics, he is the editor of* Economics, Ecology, Ethics *and the author of* Steady-State Economics *(W.H. Freeman, 1980 and 1977 respectively).*

Alvaro F. Umaña, *associate professor of environmental engineering at the University of Costa Rica, has specialized in environmental chemistry, environmental economics, and energy and resource economics. He has written on energy strategies for Costa Rica and Latin America, methane generation from fermentation wastes, and the environmental impact of geothermal energy. Currently he is Coordinator for Energy Research at the University of Costa Rica and a member of the Presidential Commission for the United Nations University for Peace.*

Kenneth J. Arrow *is Joan Kenney Professor of Economics and professor of operations research at Stanford University. An economic theorist, he has written numerous books and articles on economic theory and its applications, and in 1972 he was awarded the Nobel Prize in Economic Science.*

Robert Costanza *is an assistant professor in the Coastal Ecology Laboratory, Center for Wetland Resources, and in the Department of Marine Sciences at Louisiana State University. His interests are in systems ecology and energy analysis, and he has written on the embodied energy basis for economic-ecologic systems.*

Nicholas Georgescu-Roegen *is distinguished professor emeritus of economics at Vanderbilt University. His wide-ranging interests include economics, mathematics, statistics, and thermodynamics, and he has written numerous books and articles in these fields, including* Energy and Economic Myths *(Pergamon, 1976),* The Entropy Law and the Economic Process *and* Analytical Economics: Issues and Problems *(Harvard University Press, 1971 and 1966 respectively).*

Bruce Hannon *is associate professor and director, Energy Research Group, Office of the Vice Chancellor for Research, University of Illinois at Urbana-Champaign. He has written numerous articles and reports in his field, and currently he is editor of* Energy Systems and Policy *and associate editor of* Energy.

Garrett Hardin, *professor emeritus of human ecology at the University of California-Santa Barbara, is a specialist in microbiology, evolution, and human ecology. He has published widely in his field and is the author of* Promethean Ethics: Living with Death, Competition, and Triage *(University of Washington Press, 1980),* Managing the Commons *(with J. Baden; W.H. Freeman, 1977), and* The Limits of Altruism: An Ecologist's View of Survival *(Indiana University Press, 1977).*

Alvaro F. Umaña

Introduction

This collection of papers is an effort to gather some novel approaches to study the interaction between branches of the natural and social sciences in an integrated framework. The elements of a new "paradigm" to consider economic processes in relation to physics and biology are starting to emerge more clearly, and, in some cases, they have already been erected from solid foundations. Prior to considering the main topics included in the volume, the reader may find it helpful to briefly glance at the historical development of the ideas related to energetic and ecological economics, and to consider the changes in global conditions that bestow renewed relevance to these issues in our days.

Ecology and economics derive from the same Greek root, "oikos" meaning household. The sciences deal with the study of the laws and the interactions among our greater household, namely, the ecosphere. Ecology and economics deal with production and distribution of valuable resources among complex networks of producers and consumers. Energy and material transformations underlie all these processes, and therefore both ecology and economics must comply with the fundamental constraints imposed by thermodynamics. Nevertheless, throughout their historical development, ecology and economics have had relatively little contact. With a few notable exceptions, on the whole these sciences have shared little with respect to methodology and scope of analysis.

Recently, new global problems generated by economic and population growth, environmental degradation, and rapidly changing technologies have placed renewed attention on the relation between economics and ecology. During the second half of this century, a highly interdependent world system has consolidated and introduced qualitative changes in the human perception of our relation to the rest of nature. The development and proliferation of nuclear weapons has introduced the possibility of human destruction, as well as some probability of ending most forms of life on the planet. Modern technology has allowed for rapid increases in the utilization of energy and other resources, and has led to the creation of over a million synthetic organic compounds that did not exist on the planet a few decades ago. Anthropogenic contributions to the main cycles of the elements significantly alter the global circulation of carbon, nitrogen, phosphorus, and others. The rise in the concentration of carbon dioxide and the influence on the ozone layer by fluorocarbons are clear examples of the qualitative changes introduced by technology to ecological relations on a global scale.

The space exploration program, one of the main products of the technological progress of our era, had profound repercussions on our concept of the human position on the planet. For the first time in history, humans were able to see the Earth directly from the depths of space, and thus clearly experience its character of "spaceship" with limited and fragile resources.

In 1966, Kenneth Boulding borrowed Adlai Stevenson's "spaceship" analogy for his path-breaking article on "The Economics of the Coming Spaceship Earth." This work introduces some of the fundamental concerns of a new kind of economics subject to biospheric constraints. Boulding characterizes humans and societies as open systems with respect to energy, materials, and information input and output. These systems process energy, materials, and information in order to survive and evolve. Therefore, an important aspect of the interaction of the economy and its surrounding environment can be described in terms of energy, material, and information fluxes. Systematic organization and quantification of these inputs and outputs will be considered in a later section of this Introduction.

Boulding emphasizes that the closed earth of the future requires economic principles that are different from those of the open earth of the past in several significant aspects. Among them are a completely different attitude towards consumption, and serious revision of stock and flow concepts. In a "cowboy" economy both consumption and production are regarded as desirable without limit, and the success of the economy is measured by the rate of throughput, roughly equivalent to GNP. By contrast, in the "spacemen" economy, throughput is to be minimized to maintain a given and desired stock capable of satisfying human needs. This last idea goes clearly against the prevailing income-flow concept that still dominates economics, and starts an intense debate about the goals of economic activity and growth that still goes on, and is likely to become even more heated in the future. This set of issues is addressed at different levels by most of the contributors to this volume.

In a different article on "Ecology and Economics" in that same year, Boulding analyzes the similarities and differences between the sciences, as well as some of the constraints posed by the fact that it is no longer possible to treat the economy as if it could take resources and empty wastes into the infinite reservoir of nature. According to Boulding, two important similarities between ecology and economics are the use of general equilibrium concepts, and the developmental perspective common to both sciences. These issues have also been the subject of intense debate and are addressed by most of the contributors to the volume. For example, Goergescu-Roegen has long criticized the prevalence of equilibrium concepts in economics and has argued for an irreversible, evolutionary view developed in his paper. Evolution and development imply change through time, and therefore raise questions about the future behavior of the global economy given the constraints imposed by energy and material resources of the planet. A section of this Introduction will address the issues of evolution and economics.

Starting in 1965, Nicholas Georgescu-Roegen has carefully analyzed the economic process in terms of energy and material transformations, and laid the conceptual basis for economic models with solid grounding in thermodynamics. His work has focused on the irrevocable transformation that energy and resources undergo as they pass through the economic process, and the role of thermodynamics as "in essence a physics of economic value" (Georgescu-Roegen, this volume, p. 1). Georgescu-Roegen has long been an advocate of a qualitatively different model of economic activity, one that is not based on mechanical equilibrium analogies: "The economic process, like biological life itself, is unidirectional" (Georgescu-Roegen, 1965, p. 98). This conclusion may have not been important at other times, but it brings forth critical issues of this and future stages of human development. Economists, in general, have not yet recognized the full significance of Georgescu-Roegen's work and his true standing among economic thinkers. Future events are likely to correct this situation.

Georgescu-Roegen's thinking has developed and formalized in a book and many articles, and along with Boulding he was, and remains, at the forefront of the economists that began drawing attention to the scarcity inherent in the limitations of the ecosphere. The questions raised by them have strongly influenced the development of thinking in this area, and both contributors and editors to the present volume are highly indebted to them.

This volume contains a clear and concise summary of Georgescu-Roegen's analysis of energy and matter in the economic process, and rather than paraphrasing his arguments, the reader is directed to the first six sections of "Energy, Matter, and Economic Valuation."

Another important precursor of many issues treated in this volume is co-editor Herman Daly's "On Economics as a Life Science," published in 1968. Daly sets out to explore similarities between biology and economics given that the subject matter of both is the life process. He traces biological analogies in the history of economics, introduces his "Steady-State" analogy, and studies the human economy in evolutionary context and from an ecological perspective. Daly has developed and popularized his arguments in a book on Steady-State Economics (1977), and two edited collections: Toward a Steady-State Economy (1973), and Economics, Ecology, and Ethics (1980).

Daly's solution to the problem of how to integrate the world of commodities into the larger economy of nature by means of input-output models is presented later on in this Introduction.

The rest of the chapter is organized as follows: First, a digression on thermodynamics and the economics process is presented for readers who are not familiar with the laws of energy and their manifestations in the everyday world. The laws of thermodynamics are presented, the problem of system boundaries is addressed, and recent developments in the thermodynamics of open systems are introduced.

Second, the problem of the human economy and its material and energy fluxes to and from nature is considered, primarily by borrowing from Daly and Leontief's work.

Third, the human economy is analyzed in evolutionary context in terms of the evolution of our greater household and human interactions within it. In recent years, the implications of evolutionary theory to social and economic behavior have received some attention, and some aspects will be considerd here.

A final section points to major issues that the reader might want to look for in the papers.

A Digression on Thermodynamics and the Economic Process

One of the main hypotheses presented by the papers is that the laws of energy are necessary to understand important aspects of economic activity. Therefore it may be useful to find out what these laws are, and what they have to say about biological and economic processes.

Thermodynamics is the branch of the natural sciences that deals with the flow and transformations of energy. The science began with an analysis by the engineer Sadi Carnot, who tackled the problem of how to build the best and most efficient heat engine. The first law was clearly formulated in 1847 by von Helmholtz, the second law had been discovered earlier by Carnot himself. The laws can be briefly stated as follows:

> First Law: the principle of conservation of matter and energy states that energy can neither be created nor destroyed, only transformed. Another common formulation is, "The energy of the universe is constant."
>
> Second Law: a popular statement of this law is the principle that heat flows from hot to cold, and is connected to such common-sense observations as the fact that lakes do not freeze by themselves in the middle of the summer. There are also formal definitions such as, "The entropy of the universe is always increasing," or the equivalent formulation that the entropy of a closed thermodynamic system must also increase. These statements clearly require a definition of entropy, as well a some discussion as to the appropriate boundaries for the system under consideration.

Entropy is a measure of the unavailable energy of a system, bound energy that cannot be used to perform work. When we burn a piece of coal, heat is released and the coal turns into ashes. The first law tells us that the amount of matter and energy was conserved while the second law states that a qualitative change has taken place. At the beginning, the chemical energy stored in the coal could be used to perform useful work such as steam or power generation, but in the process of being burned, the free energy loses this quality. Ultimately it always dissipates as heat into the whole system and becomes bound energy. Thus entropy is a measure of the bound or unavailable energy of a system.

The approach we have considered thus far is known as Classical Thermodynamics. There is an alternative formulation known as Statistical Mechanics. It is based on consideration of space and velocity distributions of individual molecules. Boltzman, its founder, combined some mechanical postulates with notions of probability to define entropy as thermodynamic probability. Free energy exists in ordered structures, while bound energy exists in chaotic, disordered counterparts. In an analogous fashion, free energy is associated with low entropy and low probabilities, while bound energy is associated with high entropies and high probabilities. There has been some argument among scientists about the validity of the assumptions underlying statistical mechanics; however, modern versions of statistical and thermal physics present a unified view of classical thermodynamics and statistical mechanics.

The concept of entropy in communication theory was formulated mathematically by Claude Shannon in 1948. Shannon's definition of entropy has the same name and the same functional representation as the earlier version of entropy in statistical thermodynamics. Therefore there appear to be wide applications of the second law which lead to different interpretations. Increase in entropy can be thought of as a tendency towards higher disorder or randomness, loss of information or organization, and also movement towards more probable or disordered states. While these laws hold without exception in the universe as a whole, there are certain systems that appear to move against the continuous tide of entropic degradation. Living beings in general are examples of such systems.

Before we consider the solution to the apparent contradiction, we need to consider one more application of the second law. It follows from this law that our piece of coal can only be burned once because once the unbound energy is released it can no longer be used to perform work. Let us suppose we could reverse the process, or think of an electrochemically reversible system. In order to reverse the process we would have to provide at least an equal amount of energy as was originally present. In most processes, if they are reversible at all, a lot more energy than initially present has to be supplied. It follows that recycling of an energy source can only be a losing proposition in energy terms. Much more energy has to be consumed than is produced.

We must now return to consider the appropriate boundary and some of the properties of the system under consideration. Let us consider the earth as a whole as our system, and draw the boundary somewhere in the outer reaches of the atmosphere. Seen in such a way the earth is an open system that receives high-temperature/low-entropy radiation from the sun, cosmic rays, and some small material contributions from space. The system exchanges matter and energy with its environment. In the presence of a system like the sun-earth, where low-entropy radiation impinges upon the cooler region, great steps toward order and organization take place. The flux of solar energy is either reflected, absorbed, or used up by photosynthetic organisms to produce organic matter. This product is more structured and has lower entropy than the incoming solar energy.

Systems like this, where a strong temperature gradient leads to order and organization; or any other system feeding on low entropy from its environment, are non-equilibrium thermodynamic structures. All living organisms as well as societies are structures of this kind. The contradiction between the biological order and the increase in disorder predicted by the second law cannot be resolved as long as we try to understand living systems by means of equilibrium statistical mechanics or thermodynamics. Ilya Prigogine provided a framework to resolve the apparent contradiction. Just as isolated systems develop equilibrium structures after a sufficiently long time, open systems "out of equilibrium" are sometimes associated with what Prigogine calls "dissipative structures," which are characterized by the development of order through fluctuations. These fluctuations play an important role in forcing the system from one state to another, and can be considered stochastic in nature (Prigogine, 1976).

It is necessary to point out that these theories are recent and were developed primarily for chemical systems. Applications to living systems are beginning to appear, and considerations of societal problems still lie in the future. Nevertheless, the paradigm of non-equilibrium thermodynamics offers brilliant possibilities to provide a solid connection between "natural" and "social" sciences and to clarify to some degree the principles that underlie the mystery of living beings.

The second law of thermodynamics is special in several other ways. All known laws of physics, except for this one, are invariant under time reversal; that is, they would not change if time were to "flow backward." The second law is then the one that defines "evolutionary time," which we all experience. This law provides criteria to link events in a sequential order. If we could not establish a sequential ordering of events, the principle of causality would be inoperable because it would not be possible to tell what event preceded other ones. From a detailed treatment of the concept of Time the reader is directed to Prigogine's From Being to Becoming: Time and Complexity in the Physical Sciences (1978).

The laws of thermodynamics have a special place among our knowledge of the universe. They characterize the behavior of a wide variety of systems, ranging from chemical systems, to living beings, and information theory. They are much more closely related to the processes of life and societies than mechanics. They provide a solid basis to analyze economic activity. According to Einstein:

> A theory is more impressive the greater the simplicity of its premise is, the more different kinds of things it relates and the more extended is its area of applicability. Therefore the deep impression which classical thermodynamics made upon me. It is the only physical theory of the universal content of which I am convinced that, within the framework of the applicability of its basic concepts, it will never be overthrown.
>
> (Einstein, quoted in Schlipp, 1959)

In a similar vein, Boulding claims that we should be surprised by the permanence and wide range of applicability of thermodynamics,

> ... for after all, these are on the whole laws of change, and change is a universal phenomenon in all systems. Nature is a Heracletian flux, and we never step into the same river twice. All equilibria are temporary. Indeed, equilibrium is a fiction of the human imagination and is really unknown in the real world.
>
> (Boulding, 1976, p. 3)

Eddington, the physicist, is even more decisive about the law:

> The law of entropy increases - the Second Law of Thermodynamics - holds, I think, the supreme position among the laws of nature. If someone points out to you that your pet theory of the universe is in disagreement with Maxwell's equations - then so much the worse for Maxwell's equations. If it is found to be contradicted by observation, well, these experimentalists do bungle things up some time. But if your theory is found to be against

> the Second Law of Thermodynamics, there is nothing for it but to collapse in deepest humiliation.
> (Eddington, 1953, p. 74)

Thus far, the connection between the laws of energy and the functioning of biological and social systems have been established. Now, models of human activity that incorporate material and energy flows in and out of the environment are presented.

Energy and Material Fluxes to and from the Environment

One of the tasks of ecological economics has been to develop a methodology that explicitly recognizes the energy and material inputs required by the production of commodities, as well as the pollutants and wastes generated during the production and consumption of these goods. The methodology must comply with material and energy balances, and must account for the fact that resources flow from the environment into the economy, while pollutants and waste must end up in the air, water, or land.

Input-output economics, as adapted by Leontief (1966) from its XVIII century roots in Quesnay's "Tableau Economique," is an obvious choice capable of complying with the stated constraints. Input-output models are based on accounting identities. The technological mix used in a particular industry at a given time and place determines the amounts of all inputs required to produce a given output, and the technology is assumed to remain constant. A set of input and consumption coefficients describes the combination of goods and services required by each one of the productive sectors of a particular economy. Energy and material flows to and from the environment are easily added to standard input-output models by adding an environmental sector which can be divided into different categories according to end-use needs. Technical coefficients can be obtained to estimate the amounts of various pollutants generated during production and consumption processes. Flows of energy and matter from nature to the economy must be balanced by flows from agriculture, industry, and consumption activities into the air, water, and land; plus any materials accumulated as capital goods or inventories.

Victor (1972) has reviewed the literature on the construction of models that include interactions of economic activity and the environment, primarily those based on input-output economics. According to Victor, the first published extensions of input-output analysis to allow for economic and environmental interactions are those of Cumberland (1966) and Daly (1968).

Cumberland (1966) simply adds on columns for environmental costs and benefits to a standard input-output matrix, and very little else is said about how to evaluate these costs and benefits, or how to formalize and apply the model.

Daly's (1968) model is much more comprehensive and elaborate, and some of its main characteristics will be considered below. Daly goes further than just adding unpriced inputs from nature to the economy or unpriced outputs to the environment, and he recognizes interactions outside the purely economic realm. In fact, Daly attempts to describe purely economic interrelations, purely environmental interactions, and the relations between the economy and the environment into one comprehensive model shown in Table 1.

Table 1. Input-output matrix for human and non-human interactions

FROM	TO	
	HUMAN	NON-HUMAN
Human	(2)	(1)
Non-Human	(3)	(4)

The world is divided into human and non-human sectors, as shown in Table 1. All activities are seen as exchanges or transfers of commodities within and between the economy and the non-human sector. Quadrant (2) describes interactions that take place entirely within the human sector, the domain of conventional economics. Quadrant (4) includes interactions within the non-human sector, viewed as the transfer of ecological commodities, which Daly defines as free goods (zero price) or bads (negative prices generally not observed). This sector of ecological interrelations is the subject matter of biology, ecology in particular.

Interactions betwen the human and non-human sectors are also included in the model. Quadrant (1) shows the commodities that flow from the human (economic) sector, to the non-human (ecological) sector. These are called externalities. Commodity flows in the opposite direction are free goods taken from nature into the economy. They appear in Quadrant (3). The human and ecological sectors are thus linked by commodity flows in both directions. Production of goods implies generation of energy and material wastes transferred from the economy to the ecological sector, where they interact within the sector and affect the supply of free goods, therefore Quadrant (1) is connected to Quadrant (3) via Quadrant (4). Daly defines the technical coefficients in Quadrants (1), (3), and (4) as the "biophysical foundations of economics."

Daly's conceptual model is useful to show the interdependencies between the economic and ecological sectors, and to provide a systematic basis for cataloguing exchanges between the world of production and the natural counterpart, the biological world. The approach makes explicit the relation between the economy and its surrounding environment in terms of energy and material flows. As such, it constitutes a valuable conceptual tool to establish links between the sectors. Unfortunately, application of the model to real situations is precluded by several problems.

The complexity of the interactions in a real economy require that an enormous amount of information be collected to establish an input-output model of the human and non-human sectors. As Daly himself points out, the technical coefficients in Quadrants (1), (3), and (4) have never been measured, although they are measurable or can be calculated indirectly. Additional problems are posed by aggregation and valuation in the ecological sector.

In a standard input-output table, all the quantities in Quadrant (2) are expressed in dollar units. However, no such simplification can be made in the other sectors. Ecological relations, for example, cannot be aggregated as commodities with a monetary value. Similar problems arise in Quadrants (1) and (3). They point to the great difficulties encountered when assigning money values to nature, and aggregating the complexity of energy and material flows in the world into manageable, yet meaningful tables. Data problems are not as severe in Quadrants (1) and (3) as in Quadrant (4). The first two relate directly to industrial processes about which considerable amounts of data can be collected with relative ease. Therefore, most work has been performed in establishing more complete models of these sectors. Other related problems and applications of input-output economics to ecological processes are discussed by Victor (1972).

Although much more needs to be known and quantified before complete models can be constructed, simpler versions can be useful in decision-making related to how the economy should operate, or when carrying-capacity parameters may be exceeded given present growth patterns.

Leontief (1970, 1977) has developed models of the U.S. and world economy that focus on emissions of major pollutants and abatement activities. They can be used to estimate price effect changes in pollution abatement technology as well as for evaluating alternative government policies for regulating industrial pollution. Leontief uses pollution coefficients to relate emissions to the activity of each industry. An abatement industry is added which can eliminate a technologically determined amount of pollutants.

The model estimates emissions of some major pollutants like particulate air pollution, biochemical oxygen demand, nitrogen, phosphates and suspended solids, dissolved solids, solid waste, and pesticides. Abatement activities represented in the model consist only of removal of particulate air pollution, primary, secondary, and tertiary treatment of water pollution, and landfilling or incineration of solid waste. Even for this relatively small number of parameters considered, the difficulties encountered are many. Since statistics on pollution and abatement are very scarce, estimates of coefficients in this field are necessarily rough, and assumptions about abatement levels are somewhat arbitrary. Most pollution and abatement data available are those related to developed nations. Quantitative estimates of emissions, abatement technologies, and environmental standards are available for only a few of the developing countries. Leontief (1977) applies four abatement standards or scenarios for his world model based on per capita gross domestic product, and he derives emissions projections for different regions of the world with respect to solid waste, particulate air pollutants, and water pollution as well as cost of abatement. Leontief's work is the most comprehensive empirical attempt at world models, and constitutes a solid basis for more complete future versions.

Input-output economics is based on static abstraction in which technology is assumed to remain constant. No changes in external conditions can be included in these models. The human economy, in contrast, has experienced widespread technological change in a very short period of time, and is now subject to biospheric constraints that pose new complex problems. A dynamic, evolutionary frame of reference is needed to consider this situation.

Evolutionary Theory and Economics

Evolution stands today as the organizing principle of biology and the general theory of life (Mayr, 1978). It is associated with the idea of change over long periods of time. The scientific world view is dominated by the idea that the

Universe and all living beings have evolved through a long history going back nearly 20 billion years.

Energy and matter have evolved at different levels: physical evolution leading to the formation of atoms; chemical and geochemical processes leading to the formation of stars, the earth and life; biological evolution leading to complex living structures, and eventually humans; and finally social evolution leading to the present global condition. Although the boundaries between the levels are somewhat fuzzy at times, qualitative changes separate the levels, and each one has particular characteristics. Biological evolution is fundamentally different from cosmic (physical-chemical) evolution. It is more complicated than cosmic evolution and its products are much more complex than any non-living systems. Biological evolution occurs in a much faster time scale than cosmic evolution, and social evolution is also much faster than its biological counterpart.

Biological evolution is best defined as change in the diversity and adaptation of populations of organisms (Mayr, 1978). The first theory of evolution was postulated by Jean Baptiste Lamarck in 1809. He saw a progression in nature from the simple to the complex, and invoked four principles to explain his observations: the existence of a built-in drive for perfection in all organisms, the capability of adaptation to changing environments, frequent spontaneous generation, and inheritance of acquired traits.

Darwin had conceived in 1838 an alternate mechanism to account for evolution: natural selection. In 1858, Darwin's theory of evolution was announced to the Linnean Society in London. Two of Darwin's basic postulates were consistent with Lamarck's, namely the fact that the world is not static but is evolving, and that the process of evolution is gradual and continuous. Darwin rejected spontaneous generation and postulated that all organisms might, in fact, be traced back to a single origin of life. His fourth postulate emphasized that evolutionary change is not the product of chance or a drive to perfection, but rather a result of natural selection. This process encompasses two stages; the production of variation, and selection through survival in the struggle for existence. Variations arise as random processes while selection acts as a deterministic component. The overall process has both probabilistic and selective components. During the first half of the XX century, a synthesis was achieved that amplified Darwin's theory in light of Mendel's work and the chromosome theory of heredity, population genetics, and other contributions of modern biology. Mayr (1978) summarizes the "new synthesis" as being characterized by "the complete rejection of inheritance of acquired characters, an emphasis on the gradualness of evolution, the realization that evolutionary phenomena are population phenomena and a reaffirmation of the overwhelming importance of natural selection."

Modern reductionist approaches to evolution define natural selection as changes in gene frequencies in populations. There is intense debate about these theories, and they are attacked because they omit crucial aspects of evolution such as changes in diversity and adaptation. One of these theories, Sociobiology (Wilson, 1975), is defined as the systematic study of the biological basis of all social behavior. The theory is based on the assumption that the important thing in evolution is survival of genes rather than survival of the species or group. Thus, Wilson defines natural selection as "the process whereby certain genes gain representation in the following generations superior to that of other genes located at the same chromosome positions."

The idea of social evolution as a corollary of biological evolution was popularized nearly a century ago. There are several striking analogies between biological and social evolution. While biological evolution operates through the transmission of genetic information coded in nucleic acids, social evolution operates through the transmission of knowledge, organization and values. The mechanisms of mutation at the biological level, have counterparts in the mechanism of invention, discovery and synthesis. Technological change can be thought of as a process of "mutation" and "social selection" of alternative. Many of these ideas and relations are treated in depth by Boulding's Ecodynamics: A New Theory of Societal Evolution (1978). This work is Boulding's attempt at a grand synthesis of physical, biological, and social evolution. The task is enormous and many connections and links remain weak, yet the work is highly recommended for its valuable insights into the interacting systems that make up our world.

Social evolution is much faster than its biological counterpart. Social changes in the last two centuries have been monumental and they have brought about dramatic changes in global conditions that require novel perspectives to accommodate new conditions. Since biological evolution occurs in time scales many orders of magnitude greater than social evolution, the biological world cannot adapt to these changes. Thus, social systems themselves will have to adjust.

A fundamental difference between biological and social evolution is that the latter has Lamarckian characteristics. Human beings have the capability to evolve culturally through the transmission from generation to generation of learned information. Acquired improvements can persist and can be copied. This characteristic provides hope that creative solutions to critical human problems may appear and propagate in relatively short times.

Some aspects of evolutionary theory have been applied to economics. Simon (1980) summarizes three applications in the social sciences: the "survival of profit maximizers" argument, evolutionary models of business firm models, and the current debate about evolutionary selection of traits of "egoism" and "altruism."

The postulate of profit maximization is fundamental for the theory of the business firm, the foundation of economic theories of markets and economic equilibrium. Evolutionary arguments, related to the survival of profit maximizers have been elaborated to support this assumption. Simon (1980) analyzes the controversy and concludes that "it remains an open question whether evolutionary theory can provide a satisfactory foundation for classical theories of the firm, and this domain is likely to remain for some time an active area of theoretical and empirical research." Simon also discusses implications of evolutionary theory to business firm growth and explanations of egoistic/altruistic behavior.

Evolutionary theory is applied to economics in a more comprehensive way in this volume. The mechanistic, equilibrium perspective is abandoned in favor of an evolutionary model of economic activity. The transition from an equilibrium to an irreversible model of economics is clearly depicted in Georgescu-Roegen's Figures 1 and 2. Figure 1 shows the conventional pendulum movement between production and consumption, in which the economic process can go on forever. Figure 2 depicts Georgescu-Roegen's picture of the irreversible character of the economic process: an hourglass. The upper half represents available matter-energy, while the lower half represents unavailable matter-energy. In the case of the human economy, however, the hourglass cannot be turned upside down. We have already started to feel this fundamental entropic problem, and time will be the best judge of its validity to present conditions.

Umaña (this volume) relates this evolutionary model of economics to the thermodynamics of open systems through the "self-organization" paradigm of Prigogine.

Evolutionary theory and its relation to economics is an area of active research. Along with thermodynamics, it constitutes an important new link between the natural and social sciences.

Volume Organization

The papers in this volume address questions related to the relation between energy, ecology, and economics at many different levels. Umaña and Georgescu-Roegen focus on economic activity as a general process, and they analyze contributions to the conception of the economic process that arise from the laws of matter, energy, and biology.

Umaña provides a short historical introduction to "biophysical" perspectives in economics, describes the prevalent paradigm, and some elements of the emerging biophysical conception. An attempt is made to relate economic cycles to the properties and characteristics of "dissipative structures" through the theory of thermodynamics of non-equilibrium systems.

Georgescu-Roegen starts by considering the position of standard economics with respect to the problem of natural resources and growth. The "energetic dogma" is considered next and criticized for its neglect of matter. Georgescu-Roegen proposes his version of a Fourth Law of Thermodynamics, which can be expressed in terms of the impossibility of complete recycling. This is an extremely important point that had been neglected until now.

Problems related to how matter "wears out" have not been adequately resolved by science, among other reasons for the great difficulties posed by surface phenomena in heterogeneous materials. Georgescu-Roegen shows how mankind's entropic problem requires that we keep separate books for energy and matter. Energy cannot be considered as the ultimate raw material as the proponents of the "energetic dogma" would pretend.

Other sections of Georgescu-Roegen's paper deal with energy analysis and its relation to economic valuation. Many of the proponents of energy analysis as well as some economists have argued that energy is the fundamental source of value, and that energy equivalents are the true measure of money. The energy theory of value, rejected by Georgescu-Roegen, is addressed directly and indirectly by the papers of Georgescu-Roegen, Constanza, Hannon, and Arrow. In the Postscript, coeditor Herman Daly analyzes the different perspectives and draws conclusions with respect to the shortcomings of an energy theory of value.

Georgescu-Roegen's paper has an empirical section dealing with the technological viability of direct solar energy collection systems. The disputed viability of this technology is also considered by Daly in the Postscript.

The two papers by Georgescu-Roegen and Umaña address problems at the highest level of abstraction, having to do with basic models of the economy given present conditions as well as energy and ecological constraints. The paper by Hannon is devoted to developing a method to analyze the total energy commitment necessary to produce a unit of energy by existing and new technologies. As such, the paper is directed toward empirical applications. Hannon develops criteria for evaluation of energy alternatives and then applies them to different production and conservation technologies.

Hannon does not advocate an energy theory of value; however, he thinks that his method for calculation of energy costs of energy technologies is important for policy-making given the shortcomings of the market system for adequate energy planning. Arrow and Daly comment at length on Hannon's methodology, as well as on the questions of temporal dimension and discounting in economics. These are also crucial issues for society, very closely linked to the choice of future energy technologies. Choice of discount rates imply underlying asumptions about the future cost of energy. A positive discount rate implies that one unit of energy today can be traded off for more than one unit of energy in the future. Daly analyzes the validity of this assumption in light of Georgescu-Roegen's arguments and humankind's present stage of development.

Constanza devotes his paper to attempt an empirical confirmation of the energy theory of value by defining appropriate boundaries for energy and economic calculations. Constanza, a representative of the Odum school of energy analysis, defends the energy theory of value and argues that it provides an empirically accurate common denominator to analyze ecological and economic systems. Constanza also argues that there is no inherent conflict between the embodied energy theory of value and value theories based on utility. He seems to provide a biological validation of market economics by arguing that "from the ecological perspective markets can be viewed as an efficient energy allocation device which humans have evolved to solve the common problem facing all species-survival." Daly analyzes these and other points related to the energy theory of value in the Postscript.

Arrow's paper derives from his role as Discussant in the Symposium "Energetic and Ecological Economics," in which he was given the role of defending standard economics. The paper deals primarily with comments on Georgescu-Roegen's work on entropy and the economic process, as well as on the paper by Hannon, the problems of discounting, and the energy theory of value.

Hardin's article is important because it focuses squarely on the problems of values and ethics, which are inherent to all issues previously considered. Hardin's contrast of values for the squanderarchy and conserver societies shows the value changes that must take place in order to undergo a transition to energy and ecologically stable societies. Hardin argues for control of demand based on "mutual coercion, mutually agreed upon." Although he is the only one to explicitly address questions of value, ethical judgments are an important element present in most papers.

Acknowledgments

The author would like to acknowledge the support provided by the University of Costa Rica and the National Council for Scientific and Technological Research of Costa Rica (CONICIT) in the planning of the symposium: "Energetic and Ecological Economics," as well as in the preparation of this volume.

References

Boulding, Kenneth E., "The Economics of the Coming Spaceship Earth," Environmental Quality in a Growing Economy, Henry Jarret, ed., Johns Hopkins University Press, 1966, pp. 3-14.

Boulding, Kenneth E., "Ecology and Economics," Future Environments of North America, Darling and Milton, eds., Nathral History Press, 1966.

Boulding, Kenneth E., Ecodynamics: A New Theory of Societal Evolution, Sage View Press, 1978.

Cumberland, J. H., "A Regional Inter-Industry Model of Development Objectives," Regional Science Association Papers, 17, 65-95, 1966.

Daly, Herman, "On Economics as a Life Science," Journal of Political Economy, 76, 392-406, May/June 1968.

Daly, Herman, ed., Toward a Steady-State Economy, Freeman, 1973.

Daly, Herman, Steady-State Economics, Freeman, 1977.

Daly, Herman, ed., Economics, Ecology and Ethics, Freeman, 1980.

Eddington, A., The Nature of the Physical World, University of Michigan Press, 1958.

Einstein, A., quoted in Schlipp, Albert Einstein: Philosopher Scientist, Harper and Row, 1959.

Georgescu-Roegen, Nicholas, Analytical Economics: Issues and Problems, Harvard University Press, 1966.

Georgescu-Roegen, Nicholas, The Entropy Law and the Economic Process, Harvard University Press, 1971.

Georgescu-Roegen, Nicholas, "The Steady State and Ecological Salvation: A Thermodynamic Analysis," BioScience, 266-270, April 1977.

Georgescu-Roegen, Nicholas, "Matter Matters, Too," Prospects for Growth, K. D. Wilson, ed., New York Praeger, 1977, pp. 293-313.

Leontief, Wasily, Input-Output Economics, Oxford University Press, 1966.

Leontief, Wasily, "Environmental Repercussions and the Economic Structure: An Input-Output Approach," Review of Economics and Statistics, L II, 262-272, August 1970.

Leontief, Wasily, The Future of World Economy, Oxford University Press, 1977.

Mayr, Ernest, "Evolution," Evolution, Scientific American, Freeman, 1978.

Nicolis, G., and I. Prigogine, Self Organization in Non-Equilibrium Systems, Wiley Interscience, 1977.

Schrodinger, Erwin, What Is Life?, McMilland, 1945.

Simon, Herbert A., "Present and Future Frontiers of the Behavioral Sciences and Social Sciences," Science, 209, 72-76, July 4, 1980.

Victor, Peter A., Pollution, Economy, and the Environment, Allen and Unwind, 1972.

Alvaro F. Umaña

1. Toward a Biophysical Foundation for Economics

> ...underlying our economic manifestations are biological phenomena which we share in common with other species; and ... the laying bare and clearly formulating the relationships involved--in other words the analysis of the biophysical foundations of economics--is one of the problems coming within the program of physical biology.
>
> Alfred J. Lotka, 1924

Introduction

Economic activity, by its nature, deals with production under constraints. Otherwise the economic problem would not arise. Each society faces a different set of constraints, determined by its natural setting, resource endowment, labor organization, technology, and social structure as a whole. These "constraints" are not fixed; they are in a process of constant change and evolve along with the rest of the society. Human economies are experiencing a transition from "cowboy" to "spaceship" economies, according to Boulding's classic analogy. Thus, a different set of constraints, arising from the composition and structure of the biosphere are becoming crucial to the future of Humanity.

The fundamental problems to be addressed in this transition stem from the necessity to develop societies that are compatible with the requirements set by global ecosystems. It is no longer possible to treat the economic process as if it could maintain a continuous exchange with the infinite reservoir of nature, from which it could draw endless resources. Energy use, natural resource availability, and ecological destruction have become critical issues in the life of modern societies.

Although ecology and economics derive from the same Greek root, and both deal with the distribution of resources among a complex network of producers and consumers, they have had relatively little contact. Recently, the growth of population, material consumption, scientific knowledge, and technological capabilities have forced upon us a new set of problems that focus on the interaction among the natural and social environments. The impact that many societies have on the natural environment has increased dramatically. Indeed, our generation has an unparalleled capacity to affect the workings of global ecosystems; not only by overloading the waste-assimilative capacity of those systems, and by creating hundreds of thousands of compounds that do not occur naturally, but also by consuming the Earth's non-renewable resources at an ever increasing rate. In many cases, human capacity to produce changes in nature has surpassed the capability to understand or foresee the effects of these changes. Humans are increasingly using larger portions of the biosphere for their own benefit, without much consideration about what the effects to the whole system might be. Many isolated environments have been irreversibly altered, and there is concern that we may be approaching more serious global effects. The global increase in carbon dioxide content in the atmosphere associated with the energy use as fossil fuels and deforestation is one example. Carbon dioxide has increased by 25% and the trend is very likely to continue. Meanwhile, scientists do not know what effect this alteration will eventually have on the planet or when the effects might be felt.

Questions like this one require that we consider economic and ecologic problems at the same time, and that we develop a framework to consider these questions. In the longer run this effort may provide a basis to analyze the possible effects of alternative growth patterns on the life support systems of the planet.

Transformations of energy and matter underlie all natural and social events; in effect nothing happens in this world without some "energy potential" to explain its occurrence. Energy from the sun drives all natural cycles, and has provided conditions from which life has sprung. Through evolution, the system has developed complex paths to tap the sun's energy based primarily on photosynthesis. In turn, the energy stored by the plants has allowed for more complex and varied organisms to evolve. Today, human societies have become increasingly dependent on some of this photosynthetic storage-energy in the form of fossil fuels, which are the life line to modern industrial societies. The rise in the prices of petroleum during the last few years has had a devastating effect on the economies of most societies. This tendency is likely to continue and be accentuated during the coming decade; therefore availability and cost of energy resources will be an important determinant of what societies will be able to achieve. It has already become evident that control over energy and material resources will also have a profound effect on international relations.

Although Einstein showed that matter and energy are equivalent physical categories, under ordinary conditions on Earth, they are not easily convertible into one another. Matter assumes highly heterogeneous forms represented by the chemical elements. These are the material resources that all societies must use to function. Materials are always required to perform work, transform energy, or accomplish any productive task. Since the material requirements of modern societies have increased dramatically during the last decades, price and availability of certain materials may pose added constraints to economies in the future. In addition, humans influence the Earth's natural circulation of materials through the atmosphere, land, water, and the biosphere as a whole. Anthropogenic sources of carbon, nitrogen, sulfur, and phosphorus already have a noticeable, and sometimes alarming impact, in the natural cycling of these elements.

The laws of the transformation of energy and matter, as well as the principles of ecosystems are fundamental constraints to spaceship economies. They help delineate the boundaries to economic processes, and set the basis for a view of economic activity more closely integrated to other natural processes in the biosphere. The development of these ideas is what I have chosen to call "biophysical foundations," after Alfred J. Lotka's pioneering efforts.

The ideas presented below have arisen to deal with the problems of the late XX century and beyond. They are the work of many people in different fields and are not complete or fully organized yet. Nevertheless, they represent a new vision of economics that challenges the mechanistic conception of most modern doctrines, and provides some criteria to evaluate alternate strategies for evolution to the global society.

Historical Perspective

That ecology and economics may be related is by no means a new idea. The historical roots of this perspective can be traced as far back as the XVIII century conceptions of the Physiocrats, who believed that only agriculture was productive. William Petty's dictum that nature is the mother and labor is the father of wealth provides another early source. One could also point to the Darwinian ideas and to John Stuart Mill and the Stationary State which he derived from the classical economists, especially Malthus and Ricardo. However, more direct roots can be found in the 1920s. Alfred Marshall, one of the fathers of Neo-Classical economics believed that the Darwinian concept of natural selection is also the most important economic principle, and that "in the later stages of economics, when we are approaching nearly to the conditions of life, biological analogies are to be preferred to mechanical ones" (Marshall, 1925). In 1924 Alfred J. Lotka published his "Elements of Physical Biology," in which he developed fundamental concepts like cycling of the elements and dynamics of evolution. In 1922 the Nobel Prize-winning chemist Frederick Soddy claimed that "the starting point of Cartesian economics are the well-known laws of the conservation and transformation of energy, usually referred to as the first and second laws of thermodynamics" (Soddy, 1922). Soddy provided the first clear formulation of energy flows in the economy and the concepts of biological and geological "capital"; in energetic terms, "the plants saved, we spend."

The economist J. A. Hobson, most famous for his theories of under-consumption and imperialism noted in 1929:

> ...all serviceable organic activities consume tissue and expend energy, the biological costs of the services they render. Though this economy may not correspond in close quantitative fashion to the pleasure and pain economy or to any conscious evaluation, it must be taken as the groundwork for that conscious valuation. For most economic purposes we are well advised to prefer the organic test to any other test of welfare, bearing in mind that many organic costs do not register themselves easily or adequately in terms of conscious pain or disutility, while organic gains are not always interpretible in conscious enjoyment (Hobson, 1929).

Unfortunately, these seminal conceptions of an economic process closely integrated into the larger economy of nature did not develop very far. It was not until the nineteen-sixties when these questions began to be addressed again by some economists.

Kenneth Boulding's "Economics of the Coming Spaceship Earth," introduces the problems of the transition from "the open to the closed Earth":

> The closed economy of the future might similarly be called the "spaceman" economy, in which the earth has become a single spaceship, without unlimited reservoirs of anything, either for extraction or for pollution, and in which, therefore, man must find his place in a cyclical ecological system which is capable of continuous reproduction of material form even though it cannot escape having inputs of energy (Boulding, 1966).

The work of Nicholas Georgescu-Roegen provided a rigorous and systematic treatment of the role of matter and energy in the economic process. He declared that "the relationship between the economic process and the Entropy Law is only an aspect of a more general fact, namely, that this law is the basis for the economy of life at all levels" (Georgescu-Roegen, 1971). He developed the entropic conception of the economic process and criticizes the mechanistic conception of economics that underlies the Neo-Classical school. At the heart of this conception is the fact that "From the viewpoint of thermodynamics, matter-energy enters the economic process in a state of low entropy and comes out of it in a state of high entropy" (Georgescu-Roegen, 1976). It is necessary to point out that not only high entropy is produced in the economic process. In reality, material resources and energy--including one of the most important forms of energy, human labor--are transformed into commodities and waste. The products themselves have lower entropy at the end than at the beginning, but the overall must necessarily end up with higher entropy products than were present initially. This entropic conception of economics points to the evolutionary and irrevocable nature of the process as opposed to the circular flows and "pendulum movements between production and consumption within a completely closed system," that characterizes Neo-Classical models. Georgescu-Roegen pointed out that although Marx recognized that nature is the source of all use values, the same idea of a circular process is the basis for the diagram of simple reproduction. Neither Neo-Classical nor Marxist models of economics have a thermodynamic evolutionary basis.

In recent years, the work of Herman Daly has addressed the problem at several different levels. He has argued that the similarities betwen biology and economics, "far from being superficial, these analogies are profoundly rooted in the fact that the subject matter of biology and economics is one, viz., the life process" (Daly, 1968). Daly's work develops the initial arguments of Georgescu-Roegen and Boulding and centers on the concepts of "Steady-State Economics." The formulation recognizes the biophysical constraints that underlie all economic activity and argues for a steady-state economy as one with constant stocks of people and artifacts, maintained at some desired, sufficient levels by low rates of maintenance "throughput." Daly's main concern has been "to develop a political economics that recognizes both ecological and existential scarcity and develops propositions at a low to intermediate level of abstraction, understandable to the layman or average citizen, rather than dictated by a priesthood of "technically competent" obscurantists" (Daly, 1977). The issues related to steady-state economics have focused sharply on the problems of growth and distribution of wealth, which are at the heart of the dominant paradigm. Daly has shown simply and clearly the contradictions inherent to the science, and has provided a real service by popularizing the arguments.

One of the problems associated with attempting to use thermodynamic arguments to describe life processes is that the laws derived by classical thermodynamics apply strictly to closed systems, while organisms, just as economies, are open systems that exchange matter and energy with their environment. The work of the Nobel prize-winning chemist Ilya Prigogine has addressed the question of the thermodynamics of open systems, and the evolution of "dissipative structures" out of equilibrium through fluctuations. Although the theories are fairly recent, they have already been applied successfully to biological and social phenomena. The "order through fluctuation" paradigm provides new insights into the evolution and stability of living systems of which human economies are an example.

In summary, we are witnessing the emergence of a new paradigm. It has arisen out of new conditions created by the development of human societies. The prevalent paradigm does not offer a satisfactory way to analyze the new problems of global ecodynamics, and the science will have to develop alternative models to deal with these problems.

The Prevalent Conception

The Neo-Classical conception of economics represents the prevalent doctrine, and despite Marshall's admonition regarding biological analogies, the discipline is centered on a mechanical equilibrium model. W. Stanley Jevons, one of the founders of the school, proudly admits that their objective was to build an economic science after the models of mechanics, as he puts it, "as the mechanics of utility and self-interest" (Jevons, 1924). This view was heavily influenced by the mechanistic conceptions that characterized the physics of the beginning of the XIX century.

A good illustration of what is meant by the mechanistic epistemology of economics is the circular flow between production and consumption that is used to portray the economic process. This is shown in Figure 1, taken from Professor Samuelson's (1973) widely used textbook. Consumer dollar-votes of demand interact with business cost-supply decisions to determine what is produced. Competition is supposed to arrive at the best alternative to decide how goods are produced. For whom things are produced is supposedly determined by supply and demand interactions for productive services: wages, interest, profits, and rent.

The figure is used to show how the economy arrives at a "general equilibrium," and how the system is "stable," that is, how it returns to its initial position after perturbations. One business cycle follows another one. Even the problem of growth is analyzed in terms of movements of the economy along an equilibrium path. The process is assumed to be able to go on forever, and this is precisely what the establishment economists recommend. No less an authority than the Council of Economic Advisors to the President of the U.S. claim that, "If it is agreed that economic output is a good thing it follows by definition that there is not enough of it" (Economic Report to the President, 1971).

The diagram does not show any dependence on factors or processes outside the system, neither is any effort made to point out that the population that supplies the labor, as well as the natural resources that make up the "land" are supplied by the "infinite reservoir of nature." This assumption went largely unnoticed for a long time; however, in recent times pollution and depletion have changed many of the conditions that supported the view. As Georgescu-Roegen points out,

> The crucial point is that the economic process is not an isolated, self-sustaining process. This process cannot go on without a continuous exchange which alters the environment in a cumulative way and without being, in its turn, influenced by these alterations" (Georgescu-Roegen, 1975).

The Neo-Classical model provides us with a picture of an abstract, self-sustaining process, in which according to

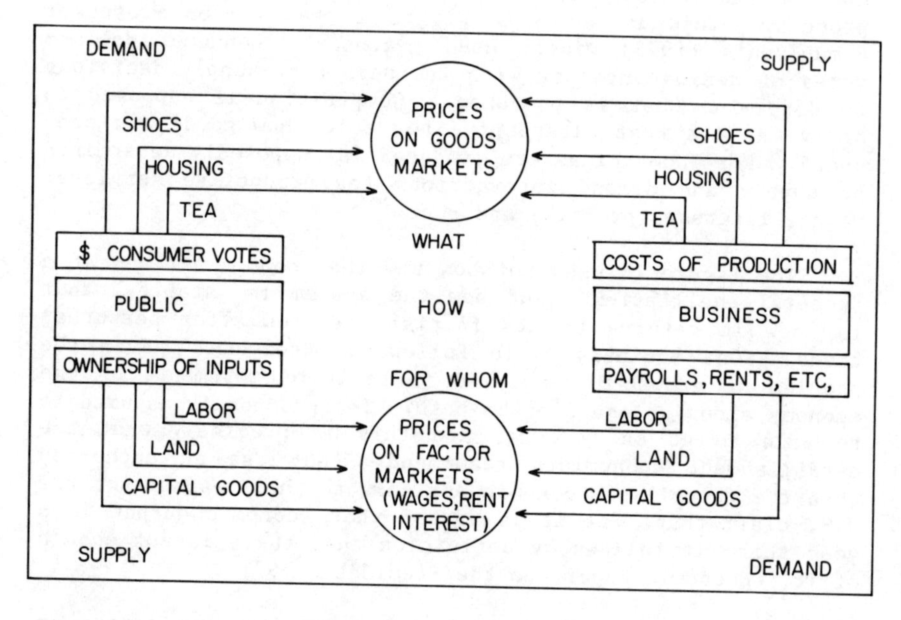

Figure 1. Diagram of the economic process. Reprinted by permission; from Paul Samuelson, Economics, McGraw-Hill, 1976.

Professor Solow, "the world can, in effect, get along without natural resources" (Solow, 1974).

The Emerging Biophysical Paradigm

Widespread concern for deteriorating environmental quality has characterized the industrialized societies during the last decade. Since environmental problems do not fall into any of the compartments of organized knowledge, scientists from many disciplines have had to study them jointly, and they have pointed to the necessity of interdisciplinary approaches to deal with questions of this kind. This approach is needed because many problems arise due to the interaction among different parts of a system. For example, human economic activity has a significant impact in the global cycling of carbon, nitrogen, phosphorus, and sulfur. In most cases economic forces are an important determinant of the types and levels of pollution, as well as the uses to which the natural environment is put.

The diagrams presented in Figure 2 attempt to illustrate the interactive view of social and economic activity with nature. It is interesting to note that the diagrams are used by four different scientists working in different fields, yet concerned about global problems. Figure 2a is taken from Stumm (1978), a chemist interested in global biogeochemical cycles and aquatic chemistry. He emphasizes this aspect in his picture, but includes energy inputs and transformations as well as economic and political elements. Figure 2b is taken from the work of the steady-state economist Herman Daly (1977). The diagram aims to show the need to maintain an adequate stock of productive assets by means of a low rate of maintenance throughput from the ecosystem. Daly argues for a balance between the damage done to ecosystems by productive activity and residual pollutants, and the benefits of the additional production, and claims that in the U.S. an adequate stock of productive assets already exists and that growth should be stopped in the interest of our long-term survival. Commoner (1976), the biologist, uses Figure 2c. Resources and energy flow from the ecosystem to the production system, where they are combined with labor and capital from the economic system. Resources are taken from the environment and pollution and heat are returned. The economic system distributes the output of production, and provides signals of how much and how to produce. The trouble, argues Commoner, is that it is the economic system that imposes requirements on the ecosystem, and not the other way around. Figure 2d is an energy diagram from the ecologist H. T. Odum (1976). The circles are external driving functions bringing in sources of potential energy. The lines are

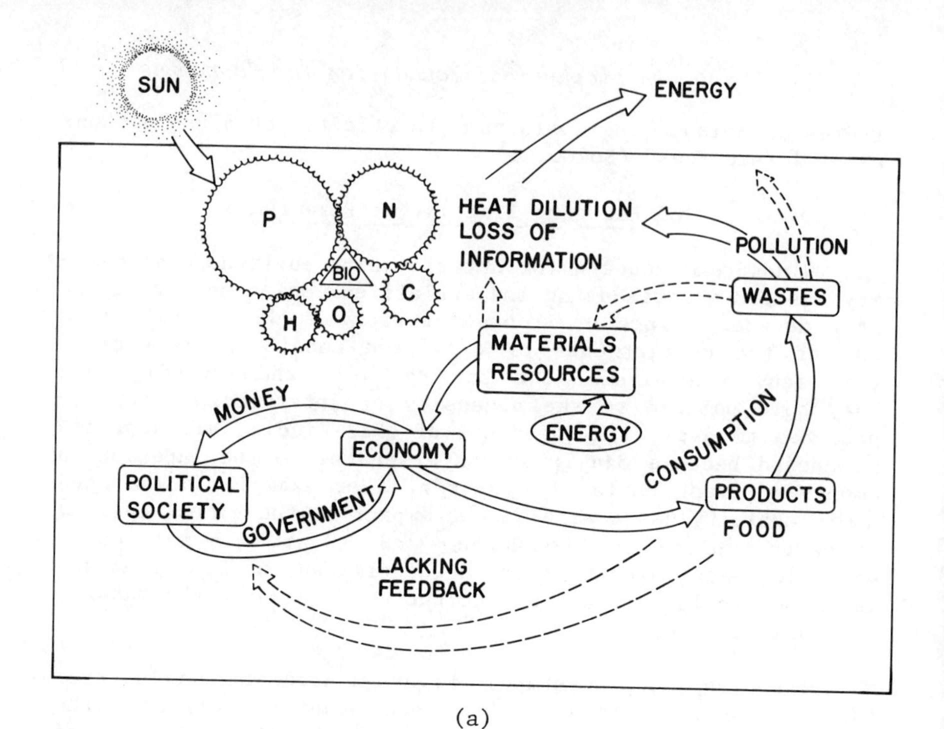

(a)

(b)

Figure 2. Interaction of social and economic activity according to (a) Werner Stumm [reprinted by permission; from Stumm, Global Chemical Cycles and Their Alterations by Man; based in part on a sketch by T.R. Blackburn; Berlin: Dahlem Konferenzen, 1977]; (b) Herman E. Daly [reprinted by permission; from Daly, Steady-State Economics: The Economics

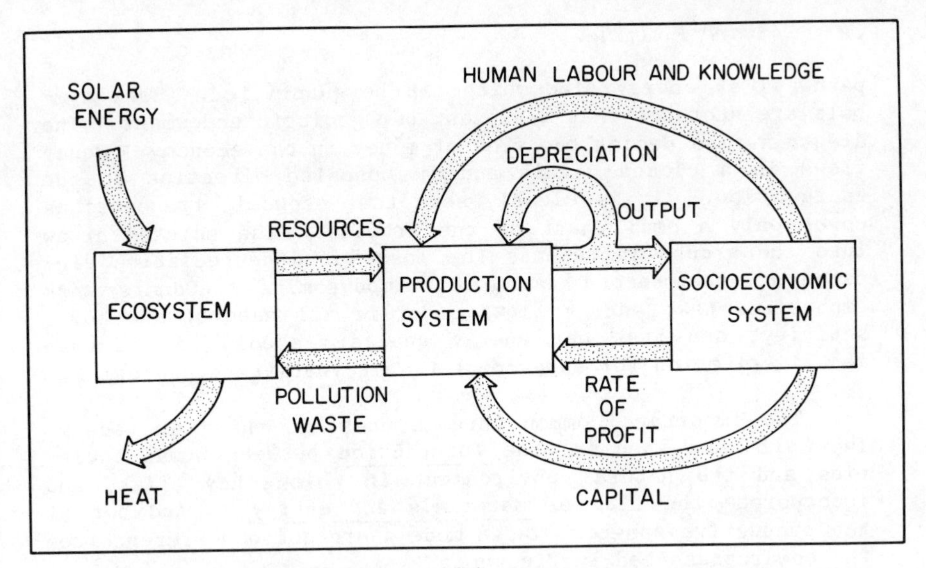

(c)

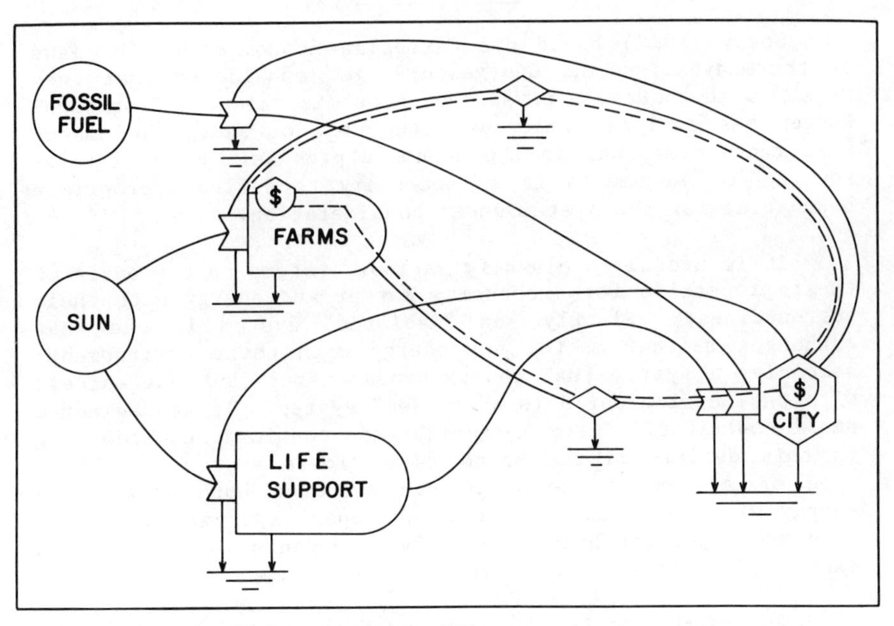

(d)

of Biophysical Equilibrium and Moral Growth, W.H. Freeman and Company, Copyright © 1977]; (c) Barry Commoner [reprinted by permission; from Commoner, The Poverty of Power: Energy and the Economic Crisis, Knopf, 1976]; and (d) Howard Odum [reprinted by permission; from Odum, Energy Basis for Man and Nature, McGraw-Hill, 1976].

pathways of energy flow which can be quantified. Tank symbols are storages that represent our geologic endowment. The diagrams also depict the flow of money in the economy. Money flows in a closed circle and in opposite direction to the energy flow. It is clearly seen that economic transactions cover only a small part of the process. The pointed arrow into the ground represents the losses and depreciation dictated by the second law of thermodynamics. Odum's work emphasizes that energy flows underlie all natural and human activity, and that net energy analysis should be a fundamental criterion for assessment of alternative technologies.

Two important common threads underlie the four models. They all make specific the interaction between human societies and the natural environment in which they exist, and incorporate the flow of materials and energy in and out of the productive sphere. These models are quite different from the one represented in Figure 1.

Thermodynamic View of the Economic Process

Soddy (1922) based his Cartesian Economics on the laws of thermodynamics and Georgescu-Roegen considered that they were "at the bottom a physics of economic value." Georgescu-Roegen has analyzed rigorously the role of energy and material transformations in the economic process. Prior to presenting his arguments it is necessary to define appropriate boundaries for the system under consideration.

It is useful to classify various systems on the basis of their properties for exchanging matter and energy with their surroundings: Firstly, an "isolated" system is one that exchanges neither matter nor energy with their environment; secondly, a system that can exchange energy (but not matter) with the outside world is a "closed" system. If we neglect a small contribution from meteorites and cosmic dust, according to this definition the Earth is a closed system. A third type of system is one which can exchange both matter and energy with the external world, an "open" system. Open systems are important because all living organisms and economies exchange matter and energy with the environment.

The Earth receives high-temperature radiation from the sun. This energy provides the thermodynamic potential for organization and structure on the planet. In the presence of a system like the sun-Earth, where low entropy radiation impinges upon a cooler region, great steps toward order and organization take place. This apparent contradiction to the laws of classical thermodynamics has been solved by science.

A new paradigm has been developed by Prigogine and others to explain the appearance of ordered structures under non-equilibrium conditions. Some details and implications of this paradigm will be considered later.

The flux of solar energy is either reflected, absorbed, or used up by photosynthetic organisms to produce organic matter. This product is more structured and has lower entropy than the incoming solar energy. Of 100 energy units coming in, less than one energy unit of organic matter is produced. The rest of the energy is absorbed as heat and radiated back into space. The solution to the contradiction is that some systems like the Earth can have a tendency toward order or organization by using low entropy generated outside the system. Therefore the Earth can maintain itself in a state of lower entropy by means of the sun's energy. A similar consideration had led Schrodinger to define life as a system that maintains its internal structure and homeostasis by feeding in low entropy from the surroundings (Schrodinger, 1945). However, if we had considered the sun inside our system, the second law tells us that the entropy of the system as a whole must have increased. The ordered structure exists at the expense of energy dissipation and waste heat in the system as a whole as to satisfy the second law.

The flux of solar energy gave rise to photosynthesis, the build-up of macromolecules, and the living order, always changing and developing, through reproduction and evolution. The organic matter generated by photosynthesis created a new form of available energy which was used to create even futher degrees of biological complexity. Through the evolutionary process, living organisms have accumulated a tremendous amount of low entropy resources that includes the biological and geological resources of the globe. In addition to the obvious example of accumulated energy in fossil fuels, more subtle mechanisms are also present. Evolution has accumulated a vast quantity of highly ordered, enormously improbable organization in the form of genetic information. The numbers of distinct and complex species, all meticulously coded in the molecular structure of DNA, represent large amounts of stored energy. Evolution has also accumulated low entropy in the extraordinary complexity and interactions among the species of the ecosystem, which represent a highly unlikely configuration of matter, equipped with mechanisms for reproduction and change.

The primary objective of economic activity is the self-preservation or reproduction of the human species. Self-preservation requires the fulfillment of certain basic biological needs, all of which require materials and low-entropy energy in different forms. All organic activities "consume tissue and expend energy, the biological requirements of the services they render," as Hobson put it. The common basis for biological or economic life is low entropy. The energy requirements of any system are Hobson's "organic costs," which he claims should be the basis for any conscious valuation.

The economic process transforms matter using low-entropy energy into products and waste. The entropic nature of economics is due to the irrevocable degradation that matter and energy undergo from available into unavailable forms. Economic activity is not a merry-go-round process, but rather one in which matter and environmental energy are irreversibly transformed.

The energetic dogma pretends to reduce all material and energy requirements, as well as economic value, into energy content alone. Georgescu-Roegen has argued that matter must be considered as a separate category, and he has formulated two laws of matter in bulk that are symmetrical to the laws of thermodynamics: no mechanical work can be performed without the use of matter in bulk, and no closed system that can exchange only energy with the outside can perform mechanical work at a constant rate indefinitely. From these laws he draws the important corollary that complete recycling is impossible.

These principles, which at first might appear too abstract, have serious repercussions for global economies when one looks at the present use of materials and energy on the planet.

Throughout history, people have learned to use many of the Earth's stocks and flows of materials and energy. The stocks are accumulated energy in the form of the planet's non-renewable resources. They are finite although not easily measurable due to technical change and other conditions of demand. One of the fundamental dilemmas of industrial societies is that they have become extremely dependent on stocks of resources. For example, the United States presently uses two units of fossil fuel energy--primarily petroleum--per unit of natural (solar, hydro, wind, tide, biomass) energy. As the stocks become depleted, the cost of obtaining energy and materials from lower-quality deposits has to rise.

During the last hundred years, technological innovations have steadily expanded the natural resource pool available to the economy by increases in efficiency, new extractive processes or techniques, and mostly plentiful energy from fossil fuels. Recovery techniques have become highly energy-intensive, and improvements in this area will be strongly dependent on energy prices. High technologies for energy generation such as fission and fusion have proven to be controversial and elusive, and industrial societies still do not have a long-term solution to the coming shortages and high prices of oil in the next two decades. Future reserves of raw material and energy sources are likely to be less concentrated or rich than the ones used in the past, and thermodynamics dictates that more energy is necessary to upgrade them. For example, the production of 50 Btu of crude oil onshore in the 1950s required about 1 Btu in production energy costs. The production of 50 Btus of synthetic crude oil from shale or coal may require about 8-10 Btus. The net energy yield is less because more feedback energy is required to process the source (Gilliland, 1977).

It is clear that the stocks of resources are bound to be exhausted some time in the future, depending on the rate at which they are used. Economists in general assume a "backstop technology," such as fusion, such that, as Solow pointed out, "at some finite cost, production can be freed of dependence on exhaustible resources altogether" (Solow, 1974). If one does not go along with this assumption, it follows that societies have to make a transition from the use of stocks to the use of flows of energy, and attempt to minimize material consumption.

The flow of energy is derived from the sun; either directly in the form of sunlight, or indirectly through wind, water power, or other more subtle mechanisms. The rate at which this energy can be used is inherently limited by the rate at which it impinges upon the earth. The amount of energy that arrives from the sun each day is enormous, several times larger than all the fossil fuels contained in the Earth. Therefore, it would appear that there is plenty of potential energy. The situation, however, is more complicated. Photosynthesis, nature's primary mechanism to capture and transform solar energy is only about one percent efficient. Hydroelectric energy, wind, and geothermal sources are highly localized and in general cannot substitute liquid fuels in transportation and other applications.

A thermodynamic analysis of the long-term prospect of economies clearly shows the present instability and points the way that societies will have to follow in the transition to renewable energy sources.

From a Mechanistic to an Evolutionary View of Economics

A thermodynamic analysis of the economic process shows the irreversible nature of the transformations of matter and energy that occur. This view, inherently different from the prevalent mechanistic conception, is evolutionary because it emphasizes the movement of the system through time and the changes that accompany the system's motion.

A new paradigm has appeared concerning the evolution of complex systems. It resolves an apparent contradiction in the interpretation of the word "evolution" in the physical and human science. In the former, for example, it has traditionally referred to the movement toward thermodynamic equilibrium and increased disorder, while in biology and the social sciences it has been associated with increasing complexity and organization. Prigogine and others have shown that a system in contact with the outside world can, under certain conditions, exhibit a quite different evolution, one corresponding to a decrease of its entropy, to a "self-organization."

A common feature of biological and social structures is that they are open systems which depend vitally on the exchange of matter and energy with their environment. However, the requirement of an open system is not a sufficient condition for "self-organization." This is only possible when the system is maintained "far from equilibrium" and if there exist certain types of "non-linear" interactions between the various components of the system. Under these conditions new structures and new types of organization can appear spontaneously, and these are termed dissipative structures. Dissipative structures link three essential factors: elementary interactions, coherent spatio-temporal organization, and fluctuations or deviations from average behavior which can generate new structures. This description contains both deterministic mechanisms and stochastic elements, the random fluctuations. These deviations from average behavior are due to the statistical nature of the system, and they are of particular importance at certain points at which new organization may emerge. Complex systems may have many bifurcation points at which fluctuations may force the system into different states or structures. Between bifurcation points, the system follows the deterministic laws, but near the points of bifurcation it is the fluctuations which play an essential role in determining the branch that the system chooses. Such a point of view introduces the concept of "history" into the explanation of the state of the system.

This type of evolution, involving determinism and chance, has been called order by fluctuation (Nicolis and Prigogine, 1977), and this extension of the physical sciences offers a new paradigm to study biological and social phenomena. The concept of "order by fluctuation" can be applied to the study of systems, whose basic units are themselves macroscopic objects containing mechanisms, their interactions with the environment and with other elements of the system. Thus, given some basic behavior patterns of the elements, their mutual interaction can lead to self-organization through successive instabilities of the collective structure.

This paradigm has been used to study the development of ecosystem populations (Prigogine, Allen, and Herman, 1977), and to model the link between employment patterns and changes in the population distribution of a region (Allen, 1978).

In economics, this paradigm may be applied to dynamic problems such as growth, fluctuations, and economic cycles. Economies are open systems that satisfy the requirements to be considered "dissipative structures." They use environmental energy to compensate internal entropy production and growth, and they exhibit the non-linear interactions necessary to develop "self-organization." The evolution of economies is characterized also by the existence of bifurcation points at which the system suffers transformation in structure and organization. Business cycle models and econometric models contain both deterministic mechanisms and stochastic fluctuations in order to describe the movement of the economy. Goodwin (1967) developed a model of cyclical growth in which the equations that describe the system are formally identical to the Volterra-Lotka predator-prey equations. Prigogine, Allen, and Herman (1977) have shown that these equations are examples of the principle of "order by fluctuation in biological systems."

Multiplier-accelerator models of the business cycle recognize that cyclical fluctuation may result from the interaction of the multiplier and the acceleration principle if one or both of these relationships is non-linear. In other models fluctuations are produced by different mechanisms, like the theory of erratic shocks in which stochastic shocks generated exogenously are imposed on the deterministic part of the solution. According to this theory, the inherent tendency of the system is to produce damped cycles. The fluctuations resulting from a single disturbance would tend to disappear over time. But further disturbances or fluctuations will in fact occur quite frequently and at random intervals (Matthews, 1967).

Non-linear relations are also sometimes considered in econometric models. Other times, different types of stochastic shocks, sometimes correlated, are imposed on the deterministic equations to model economic fluctuations and cycles (Hickman, 1978).

The "dissipative structures" paradigm opens up new alternatives to understand the dynamic properties and evolution of economies, and thus help to evaluate the changes in global constraints.

Ecological Basis for Economics

The crux of the environmental crisis is the fact that the production system is destroying the long-term reproductive capacity of global ecosystems. Thus far, the analysis has focused on energy and materials alone, but the ecosystems that support human societies are organized structures of living beings. In addition to energy and matter, the relation between humans and other biological organisms is central to this perspective. Our existence as a species is intrinsically tied to a delicate dynamic relation on the biosphere, which is increasingly subject to oversimplification and assaults by pollution and waste.

The relation between economic systems and ecosystems can be studied at different levels. Human production is always a critical intervention in nature. Each society produces the requirements of its subsistence by using resources from particular ecosystems.

The specific characteristics of the society and the natural setting give rise to unique environmental problems that can only be addressed at this particular level. More general consideration of these problems lead to questioning the role and future alternatives of humanity in the biosphere. As the level of abstraction increases, specificity is lost and assessment of impacts and prediction of future effects becomes more difficult. At the present time, knowledge of the complex systems of the biosphere is limited, and it is difficult to achieve consensus on scientific predictions about ecological threats. Prediction of long-term environmental problems are highly uncertain with respect to dynamics and severity of the threats.

The analysis of interactions in specific ecosystems and that focused on the biosphere as a whole are complementary approaches required for a fuller understanding of human intervention on natural systems. The former alternative leads to formulation of particular solutions, while the latter provides principles of general applicability. Trade and communication have contributed to the rapid evolution toward a global society throughout this century. An advanced and complex international division of production and consumption characterizes today's world. Many global environmental problems like increases in atmospheric carbon dioxide, and synthetic organics affect the entire planet and can only be properly analyzed at this level. These factors require a global perspective.

Each economic paradigm carries with it a view of the world that determines the subject of analysis and the questions considered interesting or important. Neo-Classical economics considers the individual consumer or entrepreneur as the fundamental unit of analysis, and addresses problems related to individual choice and the firm's production. In contrast to this view, the biophysical perspective focuses on the entire ecosystem and the biosphere as a whole, and addresses problems related to the interaction between natural and social environments. As the scientific capability to describe the behavior of natural systems develops, the positive aspects of the paradigm will be clarified further. This view of the world also includes inherent values and ethical positions markedly different from the dominant concepts of economics. If the Earth's ecosystems, on which all life ultimately depends, are destroyed or impaired by oversimplification and pollution, our long-term existence as a species is threatened. This fundamental proposition integrates both ethical and positive aspects of the paradigm. Although present knowledge of interactions in the biosphere is incipient, it provides some necessary elements of ecologically stable societies of the future, as well as potential strategies for the transition. The importance of energy and material constraints in the biosphere is basic and has already been considered. Ecologically stable societies will necessarily avoid irreversible changes in natural cycles, such as hydrological and biogeochemical cycles, as well as weather systems and climate. Spaceship economies must minimize the total impact on the environment. The impact that a society has on its environment is a complex function of the levels of production and consumption, technology, and population. In order to minimize this impact, societies must stabilize population, reduce consumption to minimum acceptable levels, and develop technologies that do not degrade people or nature.

The biological and geological capital of the Earth in the form of fossil fuel, minerals, and structure must be used to prepare for the transition to steady-state global economies. Humanity stands at a bifurcation point, where fluctuations in the dynamic interactions of the ecosphere will give rise to different structures and organizations. Social systems will have to provide diverse organizational forms that are flexible to respond to coming shocks and changing conditions. Diversity is the evolutionary strategy to survival.

Global and intra-national inequalities complicate the transition severely. While the basic needs of a significant portion of the human population are not met, the Earth's resources are used to satisfy trivial desires of the few. As long as severe disparities remain, ecologically stable societies are not possible because extreme inequalities are a source of conflict and material growth.

Humans may eventually achieve egalitarian societies, in harmony with each other and with the Earth; societies that will allow people to achieve their potential as human beings, whatever they may define that to be. The biophysical perspective emphasizes the fundamental value of the ecosphere as the source of life. Humans must walk the narrow edge, learning to live on Earth without threatening their life support system. This is the challenge of the present and future generations.

References

Allen, Peter M., and M. Sanglier. "Dynamic models of urban growth," J. Social Biol. Struct., 1, 165-280, 1978.

Boulding, Kenneth E. "The economics of the coming spaceship Earth," Environmental Quality in a Growing Economy, Henry Jarrett, ed., Johns Hopkins University, 1966, pp. 3-14.

Commoner, Barry. The Poverty of Power, Alfred A. Knopf, 1976.

Daly, Herman. "On economics as a life science," Journal of Political Economy, 76 (3 June 1968), 392-406.

Daly, Herman. Steady-State Economics, Freeman, 1977.

Georgescu-Roegen, Nicholas. The Entropy Law and the Economic Process, Harvard University Press, 1971.

Georgescu-Roegen, Nicholas. Energy and Economic Myths: Institutional and Analytical Economic Essays, Pergamon, 1976.

Georgescu-Roegen, Nicholas. "Energy, matter, and economic valuation," 146 Annual Meeting, AAAS, San Francisco, January 1980.

Gilliland, Martha, ed. Energy Analysis: A New Public Policy Tool, AAAS Selected Symposium Series, Westview Press, 1977.

Goodwin, R. M. "A growth cycle" in Capitalism and Economic Growth, C. H. Feinstein, ed., Cambridge Univ. Press, 1967.

Hickman, Bert. Class notes, Econ. 212 "Theory of Income and Income Fluctuations," Stanford University, Spring 1978.

Hobson, J. A. Economics and Ethics, D. C. Heath Book Co., Boston, 1929.

Jevons, Stanley. The Theory of Political Economy, London, 1924.

Lotka, Alfred J. Elements of Physical Biology, 1924. Also published as Elements of Mathematical Biology, Dover, 1956.

Marshall, Alfred. Memorials of Alfred Marshall, A. C. Pigou, ed., Macmillan & Co., London, 1925.

Matthews, R. C. O. The Business Cycle, Cambridge Univ. Press, 1967.

Nicolis, G., and I. Prigogine. Self-Organization in Nonequilibrium Systems, Wiley, 1977.

Odum, Howard, Energy Basis for Man and Nature, McGraw-Hill, 1976.

Prigogine, I., P. M. Allen, and R. Herman. "The evolution of complexity and the laws of nature" in Goals for a Global Community, Laszlo and Bierman, eds., Pergamon, 1977.

Samuelson, Paul. Economics, 9th ed., McGraw-Hill Book Co., 1973.

Schrodinger, Erwin. What Is Life, Macmillan, New York, 1945.

Soddy, Frederick. Cartesian Economics: The Bearing of Physical Science Upon State Stewardship, Hendersons, London, 1922.

Stumm, Werner, ed. Global Chemical Cycles and Their Alterations by Man, Dahlem Konferenzen, Berlin, 1977.

43-79

Nicholas Georgescu-Roegen

7220
7230
~~2260~~?
7230?

2. Energy, Matter, and Economic Valuation: Where Do We Stand?

Standard Economics vis-a-vis Natural Resources

It took the oil embargo of 1973-74 to make most, but not all, of us see the crucial role of resources of terrestrial origin for the existence of the human species. Yet a superficial reading of history would have sufficed to reveal that terrestrial resources have always been the prime motor of mankind's main actions. The Great Migration, for an eloquent example, was triggered by the soil exhaustion of the Central Asian steppes after centuries of sheep grazing. And whatever the battle cry, the deep-rooted motives of the major wars have always been the possession, at least the control, of terrestrial resources. Nothing speaks more plainly on this point than current events.

Humans raised bread and brewed beer long before they learned what causes fermentation. Similarly, people did not have to wait for science to formulate the Entropy Law in order to realize that all useful terrestrial resources are irrevocably "destroyed" through use. That the Entropy Law is the tap-root of economic scarcity should be by now a simple truth. Actually, thermodynamics is in essence the physics of economic value, as Sadi Carnat started it by his famous memoir of 1824 (Georgescu-Roegen, 1966, pp. 92-94; 1970, p. 54; 1971, pp. 6, 276-283; 1972, pp. 8-9). We must not, however, overlook the fact that scarcity stems also from the finitude of resources. It would be a mistake for an economist to argue--as Solow (1973, p. 43) does--that it is useless to take into account this finitude because it cannot "lead to any very interesting conclusions." The subject matter of economics is precisely the ways in which finite

---

This version has benefited from a careful combing by my devoted friend, Elton Hinshaw.

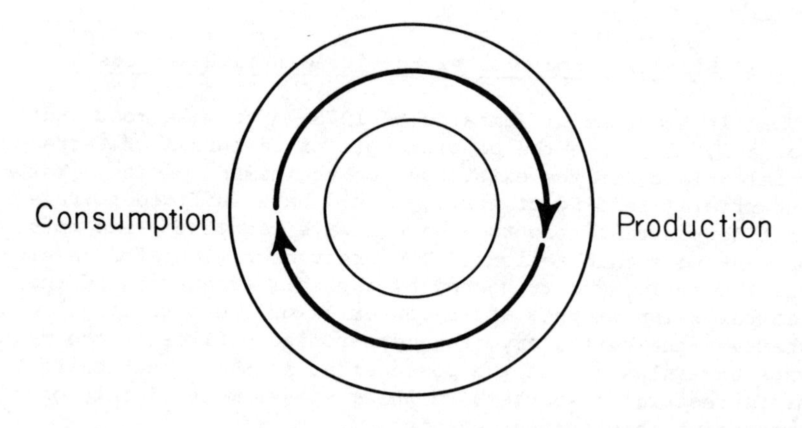

Figure 1. The standard conception of the economic process.

(i.e., scarce) means are used to satisfy given ends optimally. Take Ricardian land, which is finite. Even though it is not depletable, it has a scarcity value because it limits the yearly flow of captured solar radiation. The scarcity of terrestrial resources, however, is far more severe since their stock is not only finite, but also irrevocably exhaustible.

The ultimate reason for the paramount importance of the last scarcity is clear: ever since the human species transgressed the purely biological mode of adaptation, humans have also relied on detachable (exosomatic) organs--arrows, knives, clothes, carts, etc.--whose production requires terrestrial resources. Moreover, some of these detachable organs are now as crucial for mankind's survival as the endosomatic ones (1). In a nutshell, the exosomatic evolution is the basis of the superiority of the human species among all species. Unfortunately, it has been the source of several predicaments as well.

Standard economists have completely disregarded the unique role of natural resources in the economic process. Their vision of that process is a self-sustaining circular flow between two poles--consumption and production--as we find it represented diagrammatically in virtually all introductory manuals (Figure 1). For this truncated vision they may find an attenuating circumstance in the fact that standard economics bloomed during the fantastic mineral bonanza of the past two hundred years or so. But nothing could absolve them for the stubborn persistence in this position even after the recent developments have made us realize that the depletion of the fossil fuels, especially of the crude oil and the natural gas, is a menace to the entire world through the disintegration of the highly industrialized economies that cannot survive without an exhorbitant supply of energy (2). The defense lines are virtually invariable:

---

1. Other species also use exosomatic organs. Beavers, for example, build dams. A very interesting case is a Galapagos woodpecker finch (Cactospiza pallida), which uses a cactus spine or an appropriate twig to poke inside a wood crack for insects (Lack, 1968). But only the human species fashions exosomatic organs in a premeditated activity to use them for producing other such organs in a progressive sequence.

2. The basic vice of standard economics is a reflection of the worshipping of the mechanistic dogma by the scholarly world around the middle of the past century, that

1) Come what may, technological progress will continue just as in the past to develop exponentially so that economic growth may also follow the same unlimited rhythm.

2) Natural resources are subject to no laws different from those of any other commodity.

3) The market mechanism will also ensure the best allocation of natural resources (3).

Many a standard economist, especially those who see in the market the economic philosophical stone, strongly opposes any extra-market measure--such as the rationing of gasoline--for reacting to the present energy crisis. The objection that the inevitable rise of the free market price of gasoline (a vital commodity in the present exosomatic phase) will hurt the poor--individuals or nations--is dismissed on the ground that it concerns an issue to be resolved by modifying the prevailing income distribution. This defense is self-inflicting for the dogma that "the market always knows best." Indeed, this defense implicitly admits that the markets of the classical production factors (labor, capital, and land), which determine the income distribution, do not lead to optimality.

Unfortunately, the price complex of the standard econ-

---

is, at the time when analytical economics came to life. If that dogma has never been abandoned by economists, even after it was rejected by physics, it is only because of its capacity of grinding out mechanically one abstract mathematical exercise after another on end (Georgescu-Roegen, 1974a, 1979c).

3. See Wilfred Beckerman (1972); Carl Kaysen (1972); Robert M. Solow (1973, 1974), and J. E. Stiglitz (1979), to mention only some of the strongest defenders of the standard position. Several economists have recently recognized that as concerns monetary policies events have caught us "with our parameters down" (Walter W. Heller, 1975). Even well-known Keynesians had recently to admit that they "probably under-estimated the severity of the problem" and that events have been "more than a little disquieting" (Lawrence R. Klein, 1975; Franco Modigliani, 1979). Paul A. Samuelson (1977) also noted that "the malaise of which economists now suffer is not imaginary." However, even these witnesses did not see the other, far greater failure of standard economics, which is to ignore natural resources.

omist has spread out into a general epidemic. In approaching the problem of terrestrial resources, even natural scientists have drawn their conclusions by calculating mechanical efficiencies on the basis of the prevailing prices. Exsamples are legion, but a salient illustration of the fallacy of this procedure is supplied by a recent essay by a high authority such as Melvin Calvin (1978). After showing that the cost of a barrel of petroleum obtained from biomass was about twenty dollars at a time when the price of mineral crude oil was thirteen dollars per barrel, Calvin concluded that biomass technology will take over the field as soon as the latter price rises about twenty dollars. Yet biomass production could not turn the tables in its favor even after the OPEC price of crude oil went well above twenty dollars. The error consists of ignoring that at present all alternative sources necessitate, directly and indirectly, inputs of fossil fuels as well. The cost of obtaining energy from these sources will thus increase with the price of fossil fuels (on which more later).

The March of Ideas

Before the oil embargo the complex relationship between the special activity of human species and its supporting environment was almost a no-man's land attracting only occasionally the attention of scientists. However, one consequence of man's exosomatic existence--the qualitative deterioration of the physical surroundings--was perceived and deplored long ago. A chronicler of the seventeenth century complained that (with wood becoming increasingly scarcer) Londoners were using "sea-coale" to heat their homes, which filled London's air with "smoake" (René Dubos, 1965). It was quite natural that the deterioration of the environment, being an open surface phenomena, should strike people much earlier than the troubles of depletion. There is also the fact that, unlike depletion, environmental deterioration--especially, air and water pollution--ordinarily affects not only the future generations but also the generation that produces it.

In the United States, the Conservationist Movement has a long standing. The first move in that direction was made in 1873 by the American Association for the Advancement of Science in appointing a committee to study the problem of the preservation of forests (Samuel T. Dana, 1958). As a result, a succession of protective measures were enacted by the Congress beginning with 1891. An important sign of the growing concern with the conservation of the natural surroundings was the founding of the Sierra Club in 1892. This trend culminated in the Governors' Conference convened by President Theodore Roosevelt in 1908. On that occasion, the problem of mineral depletion also received attention, but thereafter even the problem of environmental deterioration was kept alive only by a few students, usually soil scientists concerned with the vanishing lands (e.g., R. O. Whyte and G. V. Jacks, 1939; Charles E. Kellog, 1941).

The movements of troops during World War II brought people in contact with the conditions in other countries, which rekindled the concern not only about conservation but also about the hunger affecting an overwhelming number of humans. Most of our present views of this problem go back to two signal analyses, Fairfield Osborn (1948) and William Vogt (1948). But not until the harm caused by DDT and other "wonder" chemicals struck everyone in the face did the ecological movement come to a new life. Its word was spread out in dramatic way by some works that added useful rhetoric to the new evidence (Rachel Carson, 1962; Lewis Herber, 1962; David L. Brower, 1962). This time the impetus was sufficiently strong to crystalize into some militant organizations among which the Friends of the Earth (founded in 1969) occupies a front position. The plea of preserving the "wilderness" untouched by human interference was at times so forcefully presented that one could have gotten the impression that the aim was the survival of the environment, not of the human species. Witness the fact that the day set for glorifying the faith and enhancing its public support (22 April 1970) was named Earth Day.

Accidents whose gravity could no longer be belittled not only inspired the works just mentioned, but also caused the problem of air and water pollution to become the subject of an ever increasing literary activity. To mention a few from the many, an Air Pollution Handbook containing previously published articles was already published in 1956 (P. L. Magill *et al*.). Another collection (C. Stern, 1962), which first consisted of two volumes, expanded to three volumes for the second edition (1968), and to five for the third (1976). The problem of the deterioration of the environment by the effluents of industrial activity was also one of the earliest topics entertained with a definite economic accent by some members of Resources for the Future (Allen V. Kneese, 1967; Allen V. Kneese and Blain T. Bower, 1968; Allen V. Kneese, Robert U. Ayres, and Ralph C. d'Arge, 1970).

As expected, during this early phase some ecologists began concentrating on the special topic of the ecology of the human species (Garrett Hardin, 1959; Philip Wagner, 1960;

Richard H. Wagner, 1971). The problem of population--an old preoccupation of demographers--acquired a new angle. The pressure of population on land became the favored topic of other writers as well, including a few economists (W. S. and E. S. Woytinsky, 1953; K. Sax, 1955; Joseph S. Spengler, 1959; Georgescu-Roegen, 1960; Stuart Mudd, 1964; Georg Borgstrom, 1967; Paul R. Ehrlich, 1968). Ordinarily, the debated issue was the maximum world population that could be supported by the carrying capacity of the world (if properly utilized). But Gunnar Myrdal (1940) long before perceived that the population dilemma also creates complex political problems, especially for democracies, thus foreshadowing phenomena that came to the surface only recently.

Since the point had already been reached where there are no more escape valves into undiscovered lands, most writers on the issue of population-land saw Malthusian forbodings at work. They concluded that mankind is now menaced by a "Population Bomb." To be sure, occasionally an economist also would try to approach Malthus' theory without preconceptions. But such solitary voices were quickly drawn out by prompt rebukes of the _porte-paroles_ of the establishment for whom "The Malthusian theory of population is a perfect example of metaphysics masquerading as science" (Blaug, 1968). Naturally, for economists who, like Voltaire's Dr. Pangloss, held that all is and will be well in the best of worlds, the simple mention of Malthus must be poison (4).

All this touched also the second aspect of mankind's exosomatic existence, which would operate not only if population remained constant but even if it decreased too slowly. The consequences of the eventual exhaustibility of mineral resources were considered more than a hundred years ago, albeit incidentally. In 1865, W. Stanley Jevons, just before he applied himself to constructing a new economics as "_the mechanics of utility and self-interest_" caused no little stir with his _The Coal Question_ in which he exposed the dangers awaiting the industrial power of Great Britain because of the excessive depletion of her coal reserves. But the stir soon died away (see Georgescu-Roegen, 1971). Much later, in 1923, Svante Arrhenius, a Nobelist for chemistry, considered the case of oil and arrived at the prediction that by 1950 all would be exhausted. About the same time, an astronomer,

---

4. In retrospect, Malthus was not Malthusian enough since he admitted that population may increase indefinitely provided that it does not grow faster than the means of subsistence. He did not deny, therefore, that these means can grow without limit (Georgesgu-Roegen, 1972).

Charles E. St. John, described the situation just as one would see it today: "We are using up our resources of energy in the world, and we must learn to store up the sun's energy, or to learn to get energy out of matter (by Einstein's equivalence $E = mc^2$). The world is going to be up against it some day, unless we can find out how to do some of the things in the sun--the great unexplored engine of energy" (5).

Because during World War II local shortages of some mineral resources became very stringent, a few writers became aware of mankind's most crucial predicament--its irreducible dependence on mineral resources. These writers did not need the OPEC's writing on the wall, or rather, at the service stations--"pumps closed"-- in order to perceive the problem. In England, Sir Charles Darwin endeavored to see the kind of events The Next Million Years (1953) might have in store for our species. The following year in the United States Harrison Brown tackled The Challenge of Man's Future, and a few years later, in association with James Bonner and John Weir, attempted a more modest crystal-gazing than Sir Charles Darwin's--The Next Hundred Years (1957), another contribution of everlasting value.

Apparently, the 1950's was the time when the problem of the role of energy in the existence of the human species became a lively subject, mainly becase of some inner-altar uneasiness of the Administration about incipient symptoms touching the capability of military defense. Two articles of that period by Edward Mason (1949, 1952) dealt with the issue of raw material availability for the military security. The famous Paley Report, Resources for Freedom, came out in 1952. Resources for the Future was founded the same year. The following year, Fred Cottrell and P. C. Putnam pointed out with a wealth of arguments the vital importance of energy for mankind. A few years later, Farrington Daniels (1964) dedicated a whole volume to the recipes for the direct use of solar energy, a reminiscence of St. John's older invitation.

A few economists also deserve an honorable mention because they tried their hands quite early at some analytical aspects of the problem of natural resources (John Ise, 1925;

---

5. Quoted in Thomas Nixon Carver, The Economics of Human Energy (1924). Carver's work should be mentioned mainly as a great curiosity in the economic literature. There is little in it that relates to the title. At the time, the science of economics was not sufficiently developed in the United States; nor did the topic itself present much interest.

Harold Hotelling, 1931; H. Scott Gordon, 1954). Mason (1952) addressed the more pertinent problem, that of whether economic growth can continue as in the past. Recognizing it to be a "formidable problem," he was content to foresee no dangers "over the next few decades"--a prediction that may be regarded as vindicated ex post. Undoubtedly, by limiting one's future horizon to a relatively short period one avoids coming to grips with the "formidable issue," especially at a time when a true crisis--some may say "catastrophe"--is likely to emerge just beyond such a horizon. The titles of several recent works coming from the most respected institutions betray the escapism of their authors, who boast to make quantitative predictions while limiting them usually to A.D. 2000--as if there would be no A.D. 2001, let alone A.D. 10000 (e.g., H. H. Landsberg et al., 1963; Carol Wilson, 1977; National Academy of Sciences, 1979, 1980).

Resources for the Future has constantly lived up to its name by producing works of great value, albeit biased in favor of endless, unlimited growth. Of particular importance, because of their balanced, competent analyses (especially the earlier one) are two collections edited by Henry Jarrett (1958; 1966). The Panglossian position that has characterized that institution was explicated by Harold J. Barnett and Charles Morse (1963). This contribution, which has ever since been held in the highest esteem by the economic profession, is an example of a splendidly marshalled argument supported by a false premise. The essence of its tenet is that useful mineral resources are not limited, but are continuously produced by the advances in technology that (according to the premise) will constantly happen forever. But the premise ignores that the evidence aduced pertains to the period of the unique mineral bonanza of the past two hundred years or so, which now is approaching its end.

It was not long before a few economists felt it necessary to rebel against the establishment. John Kenneth Galbraith (1958) was first (so it seems) to argue out with his usual forcefulness the necessity of reducing consumption in the face of the growing scarcity of resources. Later, independently and apparently not for the same reasons, some approached mankind's environmental predicament from a broad perspective that included objective physical arguments (Kenneth E. Boulding, 1966, 1971; John M. Gulbertson, 1971; Gerald Garvey, 1972; Georgescu-Roegen, 1965, 1966, 1970, 1971, 1972; G. Tyler Miller, 1971). There were other heretics as well. In his The Cost of Economic Growth (1967) Ezra Mishan gave a pronounced economic twist to the quality of life viewed in a broad sense well beyond the pecuniary coordinate of income. In a very attractive article, by now a classic, Garrett Hardin (1968) placed in better and also more attractive light the basic theme of Scott Gordon about the resources that belong to the entire community.

The new word spread quickly. Some ecologists took up the problem of the relations between mankind's activities and environmental resources (Barry Commoner, 1971; Howard T. Odum, 1971). Prompted mainly by ethical considerations (somewhat different from Mishan's), Herman Daly (1971,1972) revived with many new arguments the old, forgotten theme of John Stuart Mill, namely, that the stationary state necessarily implies an atmosphere of soft-pedaling and peacefulness, hence it represents the best solution to ecological salvation. Some similar thoughts guided E. F. Schumacher (1973). But the eye-catching title of his volume, _Small Is Beautiful_, has created (beyond its author's intention) an almost religious furor of admiration for what is small only. It certainly is a distorted logic not to see that size is not the criterion of human value in any direction.

Thanks to the impulse of Preston Cloud, who did the most to separate facts from fancies in the availability of minerals, the National Academy of Sciences began a study of _Resources and Man_ (1969), still a guiding source in many respects. By 1972 premonitions were strongly felt in several quarters. In January 1972, _The Ecologist_ published "Blueprint for Survival," the first cogent attempt at a broad analysis of the problems posed by the scarcity of terrestrial resources. Later in the same year, there appeared the famous report of the Club of Rome, _Limits to Growth_ by Donella H. Meadows _et al_. It had the great merit of conducting the argument with the tools so dear to the standard economist: simulation by computer and extrapolation from invariant premises (6).

What caused this awakening? Very likely, it was the intuition some had that mankind was approaching a great turning point presaged by the aggravation of the complex problems facing us which were ably outlined in a harbinger by John Platt (1969). And even though some economists appropriately issued calls for economics to be reformed from a sister science of mechanics into a human science (Herman Daly, 1968; Leonard Silk, 1972; Georgescu-Roegen, 1966, 1971, 1974b), standard economists, in general, have continued to

---

6. For an attempt at classifying the various directions from which the problems of mankind's fate has been approached, see Daly (1979).

delight in utterly empty mathematical exercises (Georgescu-Roegen, 1979c).

The Energetic Dogma

As one should have expected, after the oil embargo offered a concrete (albeit partial) example of what the "doomsdayers" were talking about, new energy experts began surging from all walks of life to expatiate to their hearts delight on the issue of natural resources and the Entropy Law (7). And as one should also have expected, this literary stampede created a great deal of confusion in the field. Unfortunately, the confusion, which explains in part the absence of any thought-out policy by the governments of the world, is not due only to the invasion of amateurism. The opinions of the most qualified authors in the field are vitiated not only by the price complex (mentioned earlier), but by still another fallacy. Unlike the former, this fallacy is the fruit of natural science. I have referred to it as the energetic dogma (Georgescu-Roegen, 1978a, 1978b, 1979a, 1979b).

"Energetics" was the first name, proposed by William Macquorn Rankine, for a discipline of a more general nature than that we now call thermodynamics. But this last word, coined by Lord Kelvin, gained general acceptance for the science now described as "mainly concerned with the transformation of heat into mechanical work and the opposite transformations of mechanical work into heat" (Fermi, 1937). Yet ex post it is clear that "energetics" would have been a far more adequate label since thermodynamics soon embraced other forms of energy (8).

---

7. Two cases suffice to show the measure of this "energy run." In a recent collection of essays, Stephen Lyons (1978), out of 22 articles only 4 are pre-1974; moreover, 13 (i.e., more than half) belong to 1977 and 1978. The distribution according to the year of publication of the works cited in a still more recent volume, Nemetz (1979), is as follows: pre-1973: 26; 1973: 18; 1974: 27; 1975: 33; 1976: 47; 1977: 96; 1978: 96; 1979: 20; forthcoming: 3. Highly symptomatic of a wide-spread academic manner in this field is that one article made no reference to any previous works.

8. Later, the term "energetics" was used to represent a position that derived the atomistic structure of matter and, hence, the kinetic interpretation of thermal energy. See Erwin N. Hiebert (1971).

The point I can hardly overemphasize is that thermodynamics has remained a science concerned only with what happens to energy alone. It has completely ignored what happens to matter, without which there can be no piston-and-cylinder performing transformations of thermal energy into mechanical work and conversely. To be sure, thermodynamics recognizes that because friction (in particular) is inherent in any natural production of mechanical work, part of the energy initially available for this effect is wasted into dissipated heat (without producing any mechanical work). No actual engine, therefore, can convert all available energy into useful work. The details of the entire story concern only various formulae for energy transformations, but nothing is said about the changes undergone by the matter of the pistons, the cylinders, the wires, the chemical solutions, or the gases themselves.

It has been said that it is unwise to struggle with a problem that a priori obstacles show to be hard to crack. It is probably because friction seems to be such a problem that it has not been incorporated into any branch of theoretical physics. Richard Feyman (1966) tells us that the laws of friction suggested by purely empirical results are found to be "'falser'" and 'falser'" under closer examination. And Ernest Rabinowicz (1965), a consummate student of that phenomenon, concludes that the subject of friction "remains a highly controversial one, and there are very few statements that can be made in this field which will find no opposition." Even the science of materials offers only tables for the force of friction between various elements, but nothing specific concerning what friction does to them.

It was thus natural that the transformations suffered by the material scaffold of machines, of life itself, and even of the surrounding reality being thus absent from the global theoretical picture, outsiders as well as physicists should come to live by the modern energetic dogma which claims that only energy matters

This belief has found additional support in the famous Einsteinian equivalence $E = mc^2$. The title of one of Alvin M. Weinberg's articles (Physics Today, 1959, p. 18) proclaims that "Energy is the Ultimate Raw Material." Hannes Alfvén (1969), a Nobel laureat in physics, also advises us that "matter, then, can be seen as a form of energy." Yet this is not the pure truth. First of all, mass is not matter in bulk, which alone is a factor of macroscopic phenomena. Also, even particles with a mass at rest (such as the proton or the neutron without which there is no matter) cannot be produced in a stable form from energy alone (9).

A highly instructive case of the issue under discussion is offered by Boulding (1966), who asserted that "there is, fortunately, no law of increasing material entropy." Very likely, Boulding was influenced by the fact that there is no formula in physics for measuring the entropic degradation of matter. The fundamental (and the only operational) formula for entropy, S, refers only to energy, specifically, to heat (which, it should be emphasized, is thermal energy in transit between a macrosystem and its environment):

$$\Delta S = Q_{rev}/T, \qquad (1)$$

where T is the absolute temperature at which the system in question receives or releases the heat Q_{rev} by a reversible process.

But the statement that expresses in the clearest way the current energetic dogma is that by Harrison Brown and his associates (1957):

> "All we need do is to add sufficient energy to the system and we can obtain whatever material we desire."

If its authors had in mind an open system (i.e., one that can exchange both energy and matter with the outside), the statement would be trivial. However, they refer it to a closed system (which can exchange only energy with the outside). Indeed, they go on to explain that the statement implies the possibility of complete recycling and, hence, of processing any ore however small may be its grade (10).

As far as I know, this view is shared without exception even by the most famous physicists. Thus, Glenn Seaborg (1972) tells us that all technical obstacles that now make for inefficiency will ultimately be eliminated by science so that we shall be able "to recycle almost any waste...to extract, transport, and return to nature when necessary all

---

9. At least not at temperatures developed after a small fraction of a second following the Big Bang. For additional details see Georgescu-Roegen (1978b, 1979a).

10. The energetic dogma permeates Harrison Brown's earlier volume (1954). But, most curiously, both volumes cite facts that go against it.

materials in an acceptable form, in an acceptable amount, and in an acceptable place so that the natural environment will remain natural and will support the continued growth and evolution of all forms of life," a vision that seems to be more grandiose than the realization of perpetual motion of the first kind.

If we accept the energetic dogma, the relationship between the economic process and the environment is represented analytically by the flow-fund matrix of Table 1 (11). To

Table 1. The Economic Process in Relation to the Environment According to the Energetic Dogma

Elements	(P_1)	(P_2)	(P_3)	(P_4)	(P_5)
			Flow Coordinates		
CE	x_{11}	$-x_{12}$	$-x_{13}$	$-x_{14}$	$-x_{15}$
K	$-x_{21}$	x_{22}	$-x_{23}$	$-x_{24}$	$-x_{25}$
C	*	*	x_{33}	*	$-x_{35}$
RM	*	$-x_{42}$	$-x_{43}$	x_{44}	*
ES	$-e_1$	*	*	*	*
W	w_1	w_2	w_3	$-w_4$	w_5
DE	d_1	d_2	d_3	d_4	d_5
			Fund Coordinates		
Capital equipment	K_1	K_2	K_3	K_4	K_5
People	H_1	H_2	H_3	H_4	H_5
Richardian land	L_1	L_2	L_3	L_4	L_5

avoid irrelevant issues, I shall divide the economic process into these consolidated processes and aggregated commodities that are relevant for the problem under consideration:

P_1: produces "controlled" energy, CE, from energy in situ, ES:

P_2: produces "capital" goods, K;

P_3: produces consumer goods, C:

11. For this mode of representing analytically a multi-process structure, see Georgescu-Roegen (1971).

P_4: recycles completely the material wastes, W, of all processes into recycled matter, RM;

P_5: maintains the population, H.

The special features of the foregoing picture must be well marked. First, neither a growing nor a declining economy can provide an acid test for the energetic dogma. Material growth cannot feed on an environmental flow of energy alone (12), whereas a declining economy may very well need no flow of environmental matter. The test case must therefore be a stationary process, or (with Marx's more felicitous terminology) a reproducible one.

But regardless of the actual system, one point is beyond any doubt ever since Lord Kelvin (1881) observed more than one hundred years ago that the energy is never lost, only part of it turns into dissipated heat and thus becomes unavailable to us. All processes of Table 1, therefore, necessarily produce dissipated (unavailable) energy, DE, which returns to the environment. The energy flows between the economic process and the environment are the input flow e_1 and the output flow $d = \Sigma d_1$. In the energetic model, however, no matter has to be brought into the economic process from the environment, and none leaves the economic process. All matter is completely recycled within that process.

Second, the representation of Table 1 reflects one elementary aspect of reality that needs unparsimonious emphasis in view of the energetic dogma and the "flow complex" that seems to dominate modern economic thought (Georgescu-Roegen, 1966; 1971). Like all actual processes, the economic process cannot exist without a material scaffold represented by its agents, the fund coordinates: capital equipment, K_i; people, H_i and H; and Ricardian land, L_i. We can never handle energy without a material lever, a material receptor, or a material transmitter. We ourselves are material structures. Biological life cannot exist without such structures. One can hardly presuppose that the energetic dogma goes so far as to claim that actual processes require no material structures of the kind we recognize side by side with energy at the macro-level.

Third, the output flow of capital, x_{22}, is destined to maintain the capital funds K_i in a reproducible condition, which means that their wear and tear is compensated for by

---

12. Obvious though this point may seem, it calls for some technical justifications that are outlined later.

the maintenance flows x_{2i}. Similarly, the flows x_{i5} maintain the population H "constant." These are the elementary conditions for the P_i's to be reproducible. And since in the case under consideration all flows must be expressed in physical units (calories or moles, for instance), the following equalities must always prevail as an aggregated translation of the conservation laws at the macro-level:

$$
\begin{aligned}
d_1 &= e_1 - x_{11}, \\
d_i &= x_{1i} \qquad (i = 2, 3, 4, 5) \\
w_1 &= x_{21}, \\
w_2 &= x_{42} - x_{22}, \\
w_3 &= x_{23} + x_{43} - x_{33}, \\
w_4 &= x_{44} - x_{24}, \\
w_5 &= x_{25} + x_{35}.
\end{aligned} \tag{2}
$$

Fourth, every recipe P_i is assumed to be feasible; that is, it can produce its product provided it is supported by the specified funds and is fed the specified inputs. However, the feasibility of every recipe P_i does not necessarily imply the viability of the technology represented by all the processes together (a crucially important point to be retained for further reference). The necessary and sufficient conditions for the viability of the technology of our reproducible economic system is given by the inequalities $x_{i5} \geq x^0_{i5}$, x^0_{i5} being a minimum determined by the "normal" standard of living, and by the well-known relations

$$
\begin{aligned}
\Sigma'_{1i} &= x_{11}, &\qquad \Sigma' x_{2i} &= x_{22}, \\
x_{35} &= x_{33}, &\qquad \Sigma' x_{4i} &= x_{44}, \\
\Sigma' w_i &= w_4,
\end{aligned} \tag{3}
$$

where the prime accent shows that the variable subscript cannot be equal to the fixed one.

Matter Matters, Too

Ever since my first thoughts on the entropic nature of the economic process, my position has been that in any system that performs work of any kind not only free energy but also "matter arranged in some definite structures" continuously and irrevocably dissipates (Georgescu-Roegen, 1965, 1966, 1970, 1971, 1972). At the time, I was under the impression that that position was common knowledge at least among natural scientists and, hence, in no need of justification. Only as I became familiar with a wider literature have I discovered that, on the contrary, the energetic dogma constitutes the general view. I then began to offer specific arguments to prove that matter matters, too, which led me to formulate a new law, to which I referred as the Fourth Law of Thermodynamics (a not very fortunate choice) (13). One formulation of this law denies the possibility of a particular system that may perform work (14):

A closed system cannot perform work indefinitely at a constant rate (15). I have referred to the system denied by this law as perpetual motion of the third kind (16). The economic process as viewed by the energetic dogma (Table 1) consists of such a motion.

As it happens with the Entropy Law, the Fourth Law is susceptible of other, equivalent formulations that are far more transparent. One such formulation states that:

---

13. I presented that law for the first time at the AAAS meeting in Boston, 21 February 1976, and expanded the arguments in its favor in several articles, to which the reader may refer for more details (Georgescu-Roegen, 1977a; 1977b; 1978a; 1978b; 1979b).

14. Actually, all thermodynamic laws can be formulated as denials. The First Law denies the possibility of obtaining work without utilization of energy (perpetual motion of the first kind). Less known is the formulation of the Entropy Law that no engine operating in cycles can perform work by using only the thermal energy of a single bath (such an engine would constitute a perpetual motion of the second kind).

15. To recall, a closed system can exchange energy (of any form) but not matter with the outside. This term is often applied to a system that can exchange neither energy nor matter with its environment. The correct terminology for such a system is "isolated."

16. To my knowledge, only Zemansky (1968) uses the same term, but he denotes by it a system from which friction, viscosity, and the like are absent. My own definition, I submit, is analytically more relevant since it does not necessarily exclude such properties of matter.

In a closed system, available matter continuously and irrevocably dissipates, thus becoming unavailable (17).

Still another one, perhaps the most relevant for the present crisis, is

Complete recycling is impossible.

The main arguments in favor of the Fourth Law invoke, first, the incontrovertible fact that material objects wear out in such a way that small particles (molecules) originally belonging to these objects are gradually dissipated beyond the possibility of being reassembled. These particles constitute unavailable matter; they can never be recycled, as is so clearly evidenced in particular by worn out automobile tires (18). What we can recycle is only what we find in the garbage and the junk yards (the garbojunk, as I have called it), that is, matter that is still available but is no longer in a useful form.

A possible objection (inspired by the statistical theory of thermodynamic phenomena) may maintain that we should be able to recycle all the molecules dissipated from a worn out penny just as we can reassemble the beads of a broken necklace in an ordinary room. This comparison extrapolates macroscopic operations to the microscopic domain, a fallacy of which quantum mechanics has convinced us all by now. Besides, what if the necklace broke somewhere in Manhattan instead of an ordinary room? To reassemble all those beads would certainly necessitate an immense effort spread over a very long time, during which other objects (used in the operation) will be worn out and hence will have to be reassembled in turn. The idea leads to a limitless

---

17. We may also say (with some concession regarding the measure of entropy of matter) that in a closed system the entropy of matter constantly and irrevocably increases. But this formulation should not be confused (as it has proved possible) with the traditional Entropy Law, which refers only to the energy in an isolated system.

18. As a refuting example of this impossibility, I have been advised that some bacteria are capable of reassembling some chemical elements from extremely weak solutions. By the same token, one may deny the existence of friction on the ground that in some circumstances it is even hard to measure it. Should we also believe in the possibility of perpetual motion of the first kind just because one can actually drive more than two thousand miles with one gallon of gasoline with a "special" automobile?

regress of the same nature as that which classical thermodynamics opposes to another idea, namely, the possibility of reversing any actual energy transformation. To recall, the point is that any attempt of this sort leaves a residual trace in the universe, a trace that can be removed by a process leaving another trace, and so on. Actually, it is surprising that still another argument of classical thermodynamics serves the same purpose in connection with the Fourth Law. To recall, for a motion to be reversible, it must proceed at an infinitesimally slow speed. Even a very small movement would then require an infinite time, which is the reason that no motion we can observe is reversible. By the same token, a complete recycling of all the molecules dissipated--say, from a given penny--if it could be achieved at all, would require an infinite time.

Thermodynamics has often been faulted for its anthropomorphic foundation: the distinction between available and unavailable energy pertains to what we humans can and cannot do. The nature of thermodynamics, however, is even more anthropomorphic than that. The impossibilities expressed by its laws are valid only because of the limits of human nature. For if our life were infinite, we would not be troubled by friction. With an infinite time at our disposal we would probably be able to recycle completely any material as well. And if we were able to move in space without limitations, we would not need a cyclic engine. With an infinite piston-and-cylinder we could convert the thermal energy of only one bath into mechanical work, as this is done in the isothermal expansion phase of the famous Carnot cycle.

As I think now of the Fourth Law, one question strikes my mind: Why has this law whose essence is so immediately obvious from everyday experience not been already incorporated into science? (19) That it has not should not surprise us greatly. Another simple phenomenon of everyday life was not made part of science until about hundred years ago, when Rudolph Clausius saw in it the essence of the Entropy Law:

---

19. To be sure, some occasional remarks reveal that some writers had some intuition of the law in the back of their minds. For example, Max Planck (1945), as he said that "diffusion, like friction and heat conduction, is an irreversible process." Because the position of H. E. Goeller and A. Weinberg (1978) is still topical (and also in great favor with economists), we may observe that they would not have promoted so strongly the feasibility of "unlimited substitutability," had they thought that recycling can be complete.

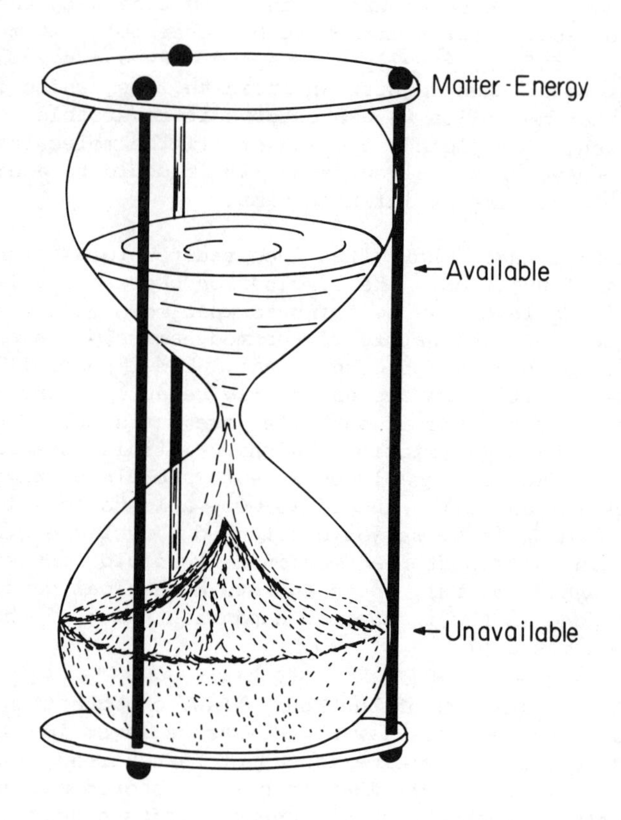

Figure 2. Thermodynamic hourglass.

Heat passes by itself from the hotter to colder body (in contact with each other), never in reverse.

One final observation on this subject. Earlier, I mentioned the insuperable difficulties yet encountered in the analysis of friction. Another and by far a more significant reason is this: Energy is a homogeneous "substance," since it can always be converted from one form into another, hence into heat. It is because of this situation that it was possible to arrive at a general formula for the degradation of energy, namely, (1). Matter, however, is not homogeneous; instead, it exists in numberless forms, each with its characteristic properties. Otherwise, the Mendeleev table would have no relevance. This being so, it seems hardly possible (at this time) to embrace the entropic degradation of matter into a single formula for all cases.

The Circulation of Energy and Matter in the Economic Process

On the basis of the foregoing analysis, we may portray in a very telling way the complete macroscopic transformation of an isolated process--possibly of the whole universe--by a thermodynamic hourglass which, like all hourglasses and unlike all cyclical clocks, marks the true passage of Time (Figure 2). The stuff of the upper half of the hourglass represents available matter-energy, which continuously pours down into the lower half and, by this very fact, becomes unavailable matter-energy. To show that this degradation is irrevocable, we shall specify that this hourglass cannot be turned upside down.

The lesson we should retain is that for mankind's entropic problem we must keep separate books, one for energy and one for matter (20). Therefore, the analytical representation of the economic process by Table 1 must be replaced by another, Table 2, in which account is taken of the fact that even a stationary economy needs a continuous inflow of environmental matter. In that table an additional process, (P_0), transforms matter in situ, MS, into controlled matter, CM. All other processes have the same roles as before and are identified by the same notations. There are, however, several important changes. First, new flows, s_i, represent the dissipated matter, DM, produced by every process and passed into the environment. Second, the recycling process,

---

20. Incidentally, to deny the foundation of this distinction would, among other things, be tantamount to denying to Ilya Prigogine the great merit of having extended the domain of thermodynamics from closed to open systems.

Table 2. The Actual Relationship Between the Economic Process and the Environment

Elements	(P_0)	(P_1)	(P_2)	(P_3)	(P_4)	(P_5)
			Flow Coordinates			
CM	x_{00}	*	$-x_{02}$	$-x_{03}$	$-x_{04}$	*
CE	$-x_{10}$	x_{11}	$-x_{12}$	$-x_{13}$	$-x_{14}$	$-x_{15}$
K	$-x_{20}$	$-x_{21}$	x_{22}	$-x_{23}$	$-x_{24}$	$-x_{25}$
C	*	*	*	x_{33}	*	$-x_{35}$
RM	*	*	$-x_{42}$	$-x_{43}$	x_{44}	*
ES	*	$-e_1$	*	*	*	*
MS	$-M_0$	*	*	*	*	*
GJ	w_0	w_1	w_2	w_3	$-w_4$	w_5
DE	d_0	d_1	d_2	d_3	d_4	d_5
DM	s_0	s_1	s_2	s_3	s_4	s_5
R	r_0	r_1	r_2	r_3	r_4	r_5

(P_4), no longer recycles all material waste. Since dissipated matter is irrecoverably lost, (P_4), can recycle only "garbojunk," denoted by GJ. Third, another inherent aspect of the economic process is represented by a flow of items that also are returned to the environment and here are labeled "refuse," R. This flow may consist in large part of available matter and available energy but in a form that is not potentially useful to us at the moment. The crushed rock of an open-pit copper mine, some urban waste, and nuclear garbage, for example, belong to this category (21).

As in the case of Table 1, the relations:

$$\Sigma' x_{0i} = x_{00}, \quad \Sigma' x_{1i} = x_{11}, \quad \Sigma' x_{2i} = x_{22}, \quad x_{35} = x_{33},$$
$$\Sigma' x_{4i} = x_{44}, \quad \Sigma' w_i = w_4 \qquad (4)$$

represent the viability of the steady state. However, because R may include both energy and matter, we can no longer write the relations for the conservation of these items separately, as in (2).

21. In order to save space, the fund coordinates have not been included in Table 2.

Energy Analysis

Ecologists and biologists have studied the energy flows moving in the environment or in the biosphere. But with the growing concern over the availability of energy for economic uses a new field of endeavor has emerged recently. Its label is "energy analysis."

The first move in that direction is due to Howard T. Odum (1973), who revived an old idea of Cottrell (1953), namely, that what counts for mankind's exosomatic existence is net energy. Odum raised this concept to a new level of importance as the criterion for technology assessment. A different approach to the same problem was developed at a 1974 meeting organized by the IFIAS (International Federation of Institutes of Advanced Study of Stockholm). The gross energy analysis, as this approach is called, claimed to represent the only logical system for that purpose. One of its representatives even charged that he had "yet to see a rigorous definition of net energy in the scientific literature" (Slesser, 1977), which is largely true. Yet the gross analysis was not free from strong criticism either (22).

The notion of net energy is at first blush extremely simple. If the energy of one ton of coal is used to mine ten tons of coal, the net gain obviously is nine tons of coal. The rub is that for mining that coal, two pounds of copper, for instance, may also be used. Shall we then say that that mining causes a negative net matter of two pounds of copper? Odum's position, never neatly outlined analytically, is that since copper is also produced with the aid of some energy, we must deduct this last energy, too, from the energy of the ten tons of coal. It is thus clear that net energy analysis is based on the energetic dogma. But so is gross energy analysis, and for exactly the same reasons.

To set the concept of net energy analysis within a clear, noncontradictory frame of reference, we may refer to Table 1 (which reflects the energetic dogma). Clearly, x_{11} cannot represent all the net energy since its production consumes x_{21} units of capital. The production of capital, in turn, needs not only energy, x_{12}, but also an input of matter, x_{42}. This shows that our search moves into a regress that will stop only after all produced commodities are included in the calculation. We are thus led to speak of a_i, the net energy equivalent of the product of P_i. The net energy itself must be measured in some unit that can only be a

22. For details on this issue, see Georgescu-Roegen, 1979a.

unit of some energy. Since the only energy involved in the framework under discussion is the controlled energy, it stands to reason that $a_1 = 1$. It should also be noted that the a_i's have nothing to do with the Einsteinian equivalence. As we shall see next, a_2, for example, represents the amount of net energy necessary to make a unit of K available to all other sectors, to the household sector as well.

The net energy, NE, and the coefficients a_i are simultaneously determined by the system

$$\begin{aligned} NE &= x_{11} - a_2 x_{21}, \\ -x_{12} &+ a_2 x_{22} - a_4 x_{42} = 0, \\ -x_{13} &- a_2 x_{23} - a_4 x_{43} + a_3 x_{33} = 0, \\ -x_{14} &- a_2 x_{24} + a_4 x_{44} = 0, \end{aligned} \tag{5}$$

which by (3) yields the national net income in net energy terms:

$$NE = x_{15} + a_2 x_{25} + a_3 x_{35} \tag{6}$$

For gross energy analysis, the same reasoning applies, with the difference that the equivalent coefficients b_i express the amount of energy <u>in situ</u>, ES, necessary to produce one unit of the corresponding commodity. This time, the unit can only be that used for e_1 (23). Because it takes more units of energy <u>in situ</u> (say, of coal) to produce one kilowatt-hour of electricity, it follows that the equivalence coefficient for the controlled energy $b_1 > 1$.

Let X denote the transposed matrix of the first four rows and four columns of Table 1; let e denote the column vector $(e_1, 0, 0, 0)$; and let b be the column vector of gross energy equivalents. The coefficients b_i are determined by the system

$$Xb = e, \tag{7}$$

which, as in the previous case, yields the net national income measured in units of ES:

$$e_1 = b_1 x_{15} + b_2 x_{25} + b_3 x_{35}. \tag{8}$$

---

23. One disturbing difficulty to the above scheme is that not all energies <u>in situ</u> have a proper measure; nuclear fuels constitute one such case.

The comparison of (5) and (7) yields

$$NE = e_1/b_1. \tag{9}$$

This relation prompts one wondering about the justification of the polemical opposition between the two schools. Of course, the a_i's can be deduced from the b_i's, but not conversely. But this does not mean that the gross energy analysis is necessarily the better approach. Each approach suits entirely different situations. The net energy analysis fits a technology based on solar radiation alone: since this energy is so abundant and for all practical considerations inexhaustible, it does not matter how much of it is used to produce one kilowatt-hour of electricity (for example). The gross energy analysis would not work in this case; for if $e_1 = 0$--as the IFIAS rules require--then (7) yields $b = 0$. Net energy analysis, on the other hand, cannot discriminate between two fossil fuels technologies. Since e_1 does not appear in (5), for net energy analysis it does not matter whether one ton or one million tons of, say, coal must be depleted to obtain one kilowatt-hour of electricity (24).

Outside the shortcomings just mentioned, both analyses make perfect sense within the energetic dogma. It stands to reason therefore that they are vitiated from that side. To say only that a_i (or b_i) units of net (or gross) energy are necessary to produce a unit of a given commodity is a dangerous half truth. The economic process requires an inflow of matter *in situ* as well. To say with David P. Brooks and P. W. Andrews (1974) that "it is preposterous to think that we can run out of matter, since the whole earth is made out of matter," is a gross inadvertence. By the same token, we could not run out of energy, since the whole earth is also full of energy. Unfortunately, the whole earth is not made of energy that is both available and *accessible*, nor is it made of available and *accessible* matter of the kinds we need for our exosomatic organs. Because of the Fourth Law, we must keep depleting the deposits of available matter even if we lived in a steady state.

In all probability, the reason we may think that matter does not matter is that the present phase of our exosomatic evolution puts a greater pressure on the supply of terrestrial energy. But in the very long run it is some material elements that are likely to become irremediably scarce. The Earth is virtually a closed system. Actually,

---

24. The net energy criterion of efficiency has become so popular that it constitutes the *official* criterion of ERDA and OTA.

some minerals already show signs of approaching the dangerous zone soon (Cloud, 1971).

Let us then turn to Table 2. Let Y denote the transposed matrix of the first five rows and five columns of Table 2; let f denote the column vector of the net gross energy equivalents (f_0, f_1, f_2, f_3, f_4) and e the column vector (0, e_1, 0, 0, 0). As before, we have

$$Yf = e, \tag{10}$$

which yields the national income measured in units of energy <u>in situ</u>,

$$e_1 = f_1x_{15} + f_2x_{25} + f_3x_{35}. \tag{11}$$

If now g denotes the column vector of <u>matter</u> equivalents (g_0, g_1, g_2, g_3, g_4), and m denotes the column vector (M_0, 0, 0, 0, 0), g is determined by the system

$$Yg = m. \tag{12}$$

From this, again we obtain the national income, this time measured in units of matter <u>in situ</u> (25).

$$M_0 = g_1x_{15} + g_2x_{25} + g_3x_{35}. \tag{13}$$

The complete truth about our tapping the natural resources is that for the production of a unit of a given commodity we must use f_i units of energy e_1 as well as g_i units of M_0.

Energy Analysis and Economic Valuation

Although my position has been, and still is, that the economic process is entropic in all its material fibers, I have explicitly rejected the idea that economic value can be reduced to a chemicophysical value through the concept of entropy. Low entropy is a necessary, but not sufficient, condition for an object to have value. I have also maintained that it would be a great mistake to think that the economic process can be represented by a vast system of thermodynamic equations. The economic process moves through

---

25. In my tables, for the sake of simplifying the notations, I assumed only one kind of energy <u>in situ</u> and one kind of matter <u>in situ</u>. The results are easily extended to the general case of several kinds of energies and matters. But one should not overlook the gigantic problem of applying the above models to actual situations.

an intricate webb of anthropomorphic categories, of utility and labor in the first place. Its true product is not a physical flow of high entropy, but the immaterial flux of the enjoyment of life obtained by the drudgery of work (Georgescu-Roegen, 1966, 1971, 1972).

The emergence of energy analysis, itself a by-product of the energy crisis, induced some energy analysts and even some economists to argue that the energy equivalents are the true measure of money. The root of the idea goes back to R. S. Berry (1972): "If the economists in the market place were to determine their estimates of shortages by looking further and further into the future, these estimates would come closer to the estimates made by their colleagues, the thermodynamicists." Energy analysts pushed the point to its furthest limit: "It makes sense to measure the cost of the things done, not in money, which is after all nothing more than a highly sophisticated value judgment, but in terms of thermodynamic potential" (Slesser, 1975). The idea was adopted and expounded with additional arguments by some economists (Gilliland, 1976; Huettner, 1976).

The error, for error there is, can be traced back to the economist's flow complex, shared also by energy analysts. To recall, the complete description of the economic process involves both flow and fund coordinates. Referring, for simplification, to Table 1, let $p = (p_1, p_2, p_3, p_4)$ be the column vector of the prices of the flow products. The funds remain constant; they provide only <u>services</u>. Let p_K, p_H, p_L be the prices of the services provided by the fund elements K, H, L during their specific periods, and let

$$B_i = p_K K_i + p_H H_i + p_L L_i. \tag{14}$$

If B denotes the column vector (B_1, B_2, B_3, B_4), any price constellation must satisfy (in principle) the system

$$Xp = B, \tag{15}$$

which yields the expression of the net national income in its two forms

$$p_1 x_{15} + p_2 x_{25} + p_3 x_{35} = \Sigma B_i. \tag{16}$$

Comparing now (15) with (5) and (7), it is perfectly clear that in absolutely no situation can the prices p_i be proportional to either the net or the gross energy equivalents. The technical coefficients, x_{ik}, completely determine the energy coefficients b_i, but they only affect prices. To claim that economic values can be reduced to "embodied energy" is a more extreme falsification of actuality than Karl Marx's theory which equates value to "embodied labor," for the latter theory assumes only that P_K and P_L are null while (15) still stands.

Matter and Technology Assessment

"Technology" is used in various senses. For the purpose of this paper, "technology" shall mean a set of feasible recipes such that one recipe exists for every produced commodity. To clarify, putting a man on the moon or heating a home with solar collectors are feasible recipes; to obtain electricity by fusion is not (yet). Technologies, in turn, can be viable or not viable depending on whether or not equations (4) are satisfied.

Let $T^1(e_1^1, M_0^1)$ and $T^2(e_1^2, M_0^2)$ be two viable technologies that yield the same net real income. Naturally, we should like to have a criterion for deciding which one is ecologically preferable. Of course, this question is relevant only if $e_1^1 < e_1^2$ and $M_0^1 > M_0^2$. If both technologies are based on some terrestrial energy, there is no way by which any energy and matter analysis alone can supply the answer. The nature of the issue is purely ecological for it involves a multitude of varying imponderables and historical uncertainties. If both T^1 and T^2 are solar technologies, then the answer is simple. Matter alone prevails; hence T^2 is to be preferred. Curiously, therefore, the answer does not call for some net analysis (which has been suggested with a solar technology in mind). It needs matter analysis.

If one technology is solar and the other terrestrial (a case that may become topical some day), the answer is again definite if T^2 is the solar technology: T^2 is the preferred one. The reverse case leads to a sequential program: use the terrestrial source until all energy *in situ* is depleted.

But the role of matter in relation to technology assessment is even greater than the preceding considerations suggest. The reason is that technologies using energies of either weak or strong intensity need a great amount of matter. In one case, for concentrating the weak flow, in the other, for containing the dangerous source. Who knows whether the fusion reactor (if feasible) would not be as big as Manhattan?

A still more important point. Nowadays we hear repeatedly that "solar technology is already here [so that] we can use it now." The truth is that only feasible recipes are here, a viable technology not. For such a technology we need a recipe that will produce the convertors of solar radiation with the aid of only the energy converted by them. Unfortunately, all solar recipes known at present are parasites of the current technologies and therefore, will cease to be applicable when their host is no longer alive (Georgescu-Roegen, 1978a, 1978b, 1978c, 1979a, 1979b) (26).

A General Overview of Mankind's Entropic Problem

The novel observations presented in the preceding sections bear directly on the present crucial exosomatic impasse of mankind, which few would deny. But to understand the true nature of this impasse, we must consider another aspect of our exosomatic evolution. Man has always striven not only to invent new exosomatic organs but also to discover more convenient means to produce them. History is punctuated by a sustained series of technical innovations. They brought man from the cave to the Moon in a few thousand years. But, surprising though the point may appear, only two innovations have had a truly crucial influence on man's technical powers.

The first was the discovery of how to master a fire. We may now regard fire as one of the most ordinary phenomenon, yet that discovery was momentous. For fire is a qualitative energy conversion, the conversion of the chemical energy of combustible materials into caloric power. Moreover, fire leads to a chain reaction: with just a small flame we can cause an entire forest, nay, all forests to burn. The mastery of fire enabled man not only to keep warm and cook his food, but also (and especially) to smelt and forge metals, and to bake bricks, ceramics, and lime. No wonder that the Ancient Greeks attributed to Prometheus (a demigod, not a mortal) the bringing of fire to man.

To the technology opened by Prometheus I, as he should be called, we may refer as the Wood Age. For centuries to come, wood served as the only source of caloric power. So that, with the industrial development growing continuously, forests began disappearing with increasing speed. During the

---

26. The above logic is perfectly proper to thermodynamic rationale. The only reason why perpetual motions are declared impossible is that no one has ever constructed one. Similarly, the present recipes do not constitute a viable technology because no one has been able to produce solar collectors without using other sources of energy.

second half of the seventeenth century, cutting forest trees had to be regulated, even restricted, both in England and on the Continent. Coal was already known as a source of caloric power, but difficulties prevented its substitution for wood on a large scale. First, coal burns dirty, a disadvantage not only for home heating but for industrial use as well. However, the greatest obstacle was the fact that any mine becomes quickly flooded, whereas the power required to drain it was not available from the sources used at that time: muscular power of humans and beasts of burden, wind, and running water (27). The impending crisis was entirely analogous to the present impasse. The technology based on wood was running out of its fuel.

Highly important for our understanding of the vicissitude of the current crisis is that neither Galilleo, nor Huygens, nor Isaac Newton himself, were able to come up with a workable solution to the crisis of that time. The solution came from another Prometheus, Prometheus II--actually, two rather common mortals, Thomas Savery and Thomas Newcomen. They invented the heat engine, which, exactly like fire, enabled man to perform a new qualitative energy conversion: the conversion of caloric power into motor power. Moreover, also like fire, the heat engine leads to a chain reaction. With just a little coal and a heat engine we can mine more coal and also other minerals from which to make several heat engines, which in turn lead to more and more heat engines.

This gift of Prometheus II represented another change of unparalleled dimensions. It enabled man to derive motor power from a new as well as more intensive source--from the fire fed by mineral fuels. We still live mainly in that technology. Using the new power to obtain more power for the satisfaction not only of legitimate needs but also of utterly absurd whims, we have caused the present bioeconomic crisis to strike us unprepared. The issue now is whether a new Prometheus will solve this crisis as Prometheus II solved that of the Wood Age.

The breeder reactor which converts fertile materials into fissionable fuels would be a third Promethean gift if its practical implementation were not still beset by greater risks and even technical snags. Indeed, doubts have recently surfaced that the physical plant of a breeder might not last long enough for plutonium to be "recycled." The sanguine

---

27. Gunpowder was already known as a source of motor power. Huygens actually thought of using it by exploding gunpowder inside a piston-and-cylinder. We know why the idea did not succeed.

hopes once surrounding controlled thermonuclear energy have almost faded away. Even Eduard Teller has recently admitted that there still is no light blinking at the end of the tunnel. It is not at all excluded that thermonuclear energy may serve only as a bomb, as is the case with gunpowder and dynamite (Georgescu-Roegen, 1972).

Undoubtedly, the situation may change suddenly and radically. But no one can be sure, one way or the other. Moreover, no one can be sure about the nature of the future Promethean gift. Nor can we command the coming of Prometheus III, as Vice-President Mondale seemed to be intent on doing recently before a Governors' Conference: "The country that developed synthetic rubber overnight during World War II [and] put the man on the Moon must now create an Apollo Project to produce alternative fuels."

The only reasonable strategy--to say "rational" would be utter intellectual arrogance--is to try to gain as great a time lead as possible to wait for the uncertain Prometheus III; alternatively, to change without great convulsions from the present high level of industrial activity to one probably analogous, but not identical, to that of the Wood Age. We must understand that our plight comes mainly from the fact that we are now the prisoners of an exosomatic structure created by an extraordinary bonanza of fossil fuels.

That we must wait has been finally recognized even by standard economists, who in great numbers and with renewed impetus now argue that nothing but the market mechanism can help us. No one could deny that a price increase normally causes a decrease in the demand. The rub is that the elasticity of demand may be such that only an exorbitant increase in price may cause a substantial decrease of demand. This is the general case for vital commodities, as petroleum products now are almost everywhere in the world. Should the price of crude oil become sufficiently high to curb its speedy depletion, only the individuals with low incomes and the poor nations would be the true "savers" of crude oil.

The market mechanism has never been able to deal with bioeconomic ills. Whenever communities have been concerned with conserving resources--forests, fish, or game--or a healthy environment, they had to introduce legal quantitative restrictions. To wit, the U.S.A. government did not apply the economists' principle "polluter pays" for diminishing the air pollution by automobiles. If the whales are now menaced by extinction, it is simply because market prices are just right for hunting them without limit.

Conservation is not the only item in a bioeconomic program, but it constitutes a crucial issue. It also is the business not of one or some nations, but of the entire human species, if homo sapiens sapiens is really sapiens.

References

Hannes Alfvén, Atom, Man, and the Universe, San Francisco: W. F. Freeman, 1969.

Harold J. Barlett and Chandler Morse, Scarcity and Growth, Baltimore: Johns Hopkins Press, 1963.

Wilfred Beckerman, "Economists, Scientists, and Environmental Catastrophe," Oxford Econ. Papers, N.S. 24 (Nov. 1972): 327-344.

R. Stephen Berry, "Recycling, Thermodynamics and Environmental Thirft," Bull. of the Atomic Scientists, 28 (May 1972): 8-15.

Mark Blaug, "Malthus, Thomas Robert," International Encyclopedia of the Social Sciences, Macmillan and Free Press, 1968, vol. 9, pp. 549-552.

"Blueprint for Survival," The Ecologist, 2 (Jan. 1972): 1-43.

Georg Borgstrom, The Hungry Planet: The Modern World at the Edge of Famine, New York: Macmillan, 1967.

Kenneth E. Boulding, "The Economics of the Coming Spaceship Earth," in Environmental Quality in a Growing Economy, Henry Jarrett, ed., Baltimore: Johns Hopkins Press, 1966, pp. 3-14.

__________, "Environment and Economics," in Environment: Resources, Pollution and Society, William W. Murdock, ed., Stanford, Conn.: Sinauer, 1971, pp. 359-367.

David L. Brower, The de facto Wilderness: What Is Its Place? San Francisco: Sierra Club, 1962.

David P. Brooks and P. W. Andrews, "Mineral Resources, Economic Growth, and World Population," Science, 185 (5 July 1974): 13-19.

Harrison Brown, The Challenge of Man's Future, New York: Viking Press, 1954.

Harrison Brown, James Bonner, and John Weir, The Next Hundred Years, New York: Viking Press, 1957.

Melvin Calvin, "Chemistry, Population, Resources," Interdisciplinary Science Reviews, 3 (Sept. 1978): 233-243.

Rachel L. Carson, The Silent Spring, Boston: Houghton Mifflin, 1962.

Preston Cloud, ed., Resources and Man, San Francisco: W. H. Freeman, 1969.

__________, "Mineral Resources in Fact and Fancy," in Environment: Resources, Pollution and Society, William W. Murdock, ed., Stanford, Conn.: Sinauer, 1971, pp. 71-88.

Barry Commoner, The Closing Circle: Nature, Man and Technology, New York: Knopf, 1971.

Fred Cottrell, Energy and Society, New York: McGraw-Hill, 1953.

John M. Culbertson, Economic Development: An Ecological Approach, New York: Knopf, 1971.

Herman Daly, "On the Economics as a Life Science," J. of Polit. Econ., 76 (May/June 1968): 392-406.

__________, The Stationary State Economy, Distinguished Lecture Series No. 2, University of Alabama, 1971.

__________, "In Defense of a Steady-State Economy," American Journal of Agricultural Economics, 54 (Dec. 1972): 945-954.

__________, "Entropy, Growth, and the Political Economy," in Scarcity and Growth Reconsidered, V. Kerry Smith, ed., Baltimore: Johns Hopkins Press, 1979, pp. 67-94.

Samuel T. Dana, "Pioneers and Principles," in Henry Jarrett (1958): pp. 24-33.

Farrington Daniel, Direct Use of the Sun's Energy, New Haven: Yale Univ. Press, 1964.

René Dubos, Man Adapting, New Haven: Yale Univ. Press, 1965. 1965.

Paul R. Ehrlich, The Population Bomb, 2nd ed., New York: Ballantine, 1971.

Enrico Fermi, Thermodynamics, Inglewood, N.J.: Prentice-Hall, 1937.

Richard P. Feynman, R. B. Leighton, and M. Sands, The Feynman Lectures on Physics, vol. I, Reading, Mass.: Addison-Wesley, 1966.

John Kenneth Galbraith, "How Much Should a Country Consume," in H. Jarrett (1958), pp. 89-99.

Gerald Garvey, Energy, Ecology, Economy, New York: W. W. Norton, 1972.

Nicholas Georgescu-Roegen, "Economic Theory and Agrarian Economics," Oxford Econ. Papers, N.S. 12 (Feb. 1960): 1-40. Reprinted as chapter 6 in Nicholas Georgescu-Roegen (1966).

__________, "Process in Farming Versus Process in Manufacturing," in Economic Problems of Agriculture in Industrial Societies, Proceedings of a Conference held by Int. Econ. Asso., Rome, September 1965, U. Papi and C. Nunn, eds., London: Macmillan, 1969, pp. 497-528. Reprinted in Nicholas Georgescu-Roegen (1976b).

__________, Analytical Economics: Issues and Problems, Cambridge, Mass.: Harvard Univ. Press, 1966.

__________, The Entropy Law and the Economic Problem, Distinguished Lecture Series No. 1, 3 Dec. 1970, University of Alabama. Reprinted in Nicholas Georgescu-Roegen (1976b).

__________, The Entropy Law and the Economic Process, Harvard Univ. Press, 1971.

__________, "Energy and Economic Myths," Lecture in the

Series Limits to Growth: The Equilibrium State and Human Society, Yale University, 8 Nov. 1972. Reprinted in Nicholas Georgescu-Roegen (1976b).

_________(a), "Mechanistic Dogma and Economics," Methodology and Science, 7 (Sept. 1974): 174-184.

_________(b), "Toward a Human Economics," Am. Econ. Rev., 64 (May 1974): 449-450.

_________(a), "A Different Economic Perspective," Paper read at the Boston Meeting of the AAAS, 21 Feb. 1976.

_________(b), Energy and Economic Myths: Institutional and Analytical Economic Essays, New York: Pergamon, 1976.

_________(a), "The Steady State and Ecological Salvation: A Thermodynamic Analysis," BioScience, 27 (Apr. 1977): 266-270.

_________(b), "Matter Matters, Too," in Prospects for Growth, K. D. Wilson, ed., New York: Praeger, 1977, pp. 293-313.

_________(a), "Myths About Energy and Matter," Lecture presented at the Conference of Energy, 27 April 1978, University of Kentucky, Growth and Change, 10 (Jan. 1979): 16-23.

_________(b), "Matter: A Resource Ignored by Thermodynamics," in Future Sources of Organic Raw Materials: CHEMRAWN I, Invited Lectures presented at the World Conference, Toronto, 10-13 July, 1978, L. E. St-Pierre and G. R. Brown, eds., New York: Pergamon, 1980, pp. 79-87.

_________(c), "Technology Assessment: The Case of the Direct Use of Solar Energy," Atlantic Econ. J., 6 (Dec. 1978): 15-21.

_________(a), "Energy Analysis and Economic Valuation," Southern Econ. J., 45 (Apr. 1979): 1023-1058.

_________(b), "Energy and Matter in Mankind's Technological Circuit," in Nemetz (1979), pp. 107-127.

_________(c), "Methods in Economic Science," J. of Econ. Issues, 13 (June 1979): 317-327.

Martha W. Gilliland, "Energy Analysis and Public Policy," Science, 189 (26 Sept. 1975): 1051-1056, and 192 (2 Apr. 1976): 12.

_________, ed., Energy Analysis: A New Public Policy Tool, Boulder, Colo.: Westview Press, 1978.

H. E. Goeller and A. M. Weinberg, "The Age of Substitutability," Am. Econ. Rev., 68 (Dec. 1978): 1-11.

H. Scott Gordon, "The Economic Theory of a Common-Property Resource: The Fishery," J. of Polit. Econ., 42 (Apr. 1954): 124-142.

Garrett Hardin, Nature and Man's Fate, New York: Holt, Rinehart and Winston, 1959.

_________, "The Tragedy of the Commons," Science, 162 (13 Dec. 1968): 1243-48.

Walter W. Heller, "What's Right With Economics," (Presidential Address), Am. Econ. Rev., 65 (Mar. 1975): 1-26.

Lewis Herber, Our Synthetic Environment, New York: Knopf, 1962.

Erwin N. Hiebert, "The Energetic Controversy and the New Thermodynamics," in Perspectives in the History of Science and Technology, H. D. Roller, ed., Norman: Univ. of Oklahoma Press, 1971, pp. 67-86.

Harold Hotelling, "The Economics of Exhaustible Resources," J. of Polit. Econ., 39 (Apr. 1931): 137-175.

David A. Huettner, "Net Energy Analysis: An Economic Assessment," Science, 192 (9 Apr. 1976): 101-104.

John Ise, "Theory of Value as Applied to Natural Resources," Am. Econ. Rev., 15 (June 1925: 284-291.

Henry Jarrett, ed., Perspectives on Conservation, Baltimore: Johns Hopkins Press, 1958.

__________, ed., Environmental Quality in a Growing Economy, Baltimore: Johns Hopkins Press, 1966.

Carl Kaysen, "The Computer That Printed Out W*O*L*F*," Foreign Affairs, 50 (July 1972): 660-68.

Charles E. Kellog, The Soil That Supports Us, New York: Macmillan, 1941.

Lord Kelvin (Sir William Thomson), Mathematical and Physical Papers, Vol. I, Cambridge, Eng.: Univ. Press, 1881.

Lawrence R. Klein, "Intractability of Inflation," Methodology and Science, 7 (Sept. 1974): 156-173.

Allen V. Kneese, "Economics and the Quality of the Environment," in Cost of Air Pollution, M. Garnsey and J. Hibbs, eds., New York: Praeger, 1967.

__________ and Blair T. Bower, Managing Water Quality: Economics, Technology, Institutions, Baltimore: Johns Hopkins Press, 1968.

__________, Robert U. Ayres, Ralph C. d'Arge, Economics and the Environment, Baltimore: Johns Hopkins Press, 1970.

David Lack, Darwin's Finches, Glouster, Mass.: Peter Smith, 1968.

Hans H. Landsberg, Leonard L. Fischman, and Joseph L. Fisher, Resources in America's Future: Patterns of Requirements and Availabilities, 1960-2000, Baltimore: Johns Hopkins Press, 1963.

Stephen Lyons, ed., Sun! A Handbook for the Solar Decade, San Francisco: Friends of the Earth, 1978.

P. L. Magill, F. R. Holden, and C. Ackley, eds., Air Pollution Handbook, New York, 1956.

Edward S. Mason, "American Security and Access to Raw Materials," World Politics, 1 (Jan. 1949): 147-160.

__________, "Raw Materials, Rearmament, and Economic Development," Quart. J. of Econ., 66 (Aug. 1952): 327-341.

Donella H. Meadows et al., The Limits to Growth, New York: Universe Books, 1972.

G. Tyler Miller, Energetics, Kinetics, and Life: An Ecological Approach, Belmont, Cal.: Wadsworth, 1971.
Ezra Mishan, The Cost of Economic Growth, London: Staples, 1967.
Franco Modigliani, "Has the Economist Lost Control of the Economy?" Bull. of the Am. Acad. of Arts and Sciences, 32 (Mar. 1979): 38-49.
Stuart Mudd, ed., The Population Crisis and the Use of World Resources, Bloomington: Indiana Univ. Press, 1964.
Gunnar Myrdal, Population: A Problem for Democracy, Cambridge, Mass.: Harvard Univ. Press, 1940.
National Academy of Sciences, Science and Technology: A Five-Year Outlook, San Francisco: W. H. Freeman, 1979.
__________, Energy in Transition 1985-2010, San Francisco: W. H. Freeman, 1980.
Peter N. Nemetz, ed., Energy Policy: The Global Challenge, Toronto: Butterworth, 1979.
Howard T. Odum, Environment, Power and Society, New York: Wiley-Interscience, 1971.
__________, "Energy, Ecology, and Economics," Ambio, 2 (1973, No. 6): 220-27.
Fairfield Osborn, Our Plundered Planet, Boston: Little Brown, 1948.
Max Planck, Treatise on Thermodynamics, 7th ed., New York: Dover, 1945.
John Platt, "What We Must Do," Science, 166 (28 Nov. 1969): 1115-121.
P. C. Putnam, Energy in the Future, New York: Van Nostrand, 1953.
Ernest Rabinowicz, Friction and Wear of Materials, New York: Wiley, 1965.
Paul A. Samuelson, "Les économistes ne sont plus ce qu'ils étaient," Le Nouvel Economiste (31 Oct. 1977).
K. Sax, Standing Room Only: The Challenge of Overpopulation, Boston: Beacon, 1955.
E. F. Schumacher, Small Is Beautiful: Economics As If People Mattered, New York: Harper and Row, 1973.
Glenn T. Seaborg, "The Erehwon Machine: Possibilities for Reconciling Goals by Way of New Technology," in Energy, Economic Growth and the Environment, Sam H. Schurr, ed., Baltimore: Johns Hopkins Press, 1972, pp. 125-138.
Leonard Silk, "Wanted: A More Human, Less Dismal Science," Saturday Rev., 55 (22 Jan. 1972): 34-35.
Malcom Slesser, "Accounting for Energy," Nature, 254 (20 Mar. 1975): 170-72.
__________, "Energy Analysis," Science, 196 (15 Apr. 1977): 259-260.
Robert M. Solow, "Is the End of the World at Hand?" Challenge (Mar.-Apr. 1973): 39-50.

_________, "The Economics of Resources and the Resources of Economics," Richard T. Ely Lecture, Am. Econ. Rev., 64 (May 1974): 1-14.

Joseph J. Spengler, "Economics and Demography," in The Study of Population, Philip M. Hauser and Otis D. Duncan, eds., Chicago: Univ. of Chicago Press, 1959, pp. 791-831.

Arthur C. Stern, ed., Air Pollution, 2 vols., New York: Academic Press, 1962.

J. E. Stiglitz, "A Neoclassical Analysis of the Economics of Natural Resources," in Scarcity and Growth Reconsidered, V. Kerry Smith, ed., Johns Hopkins Press, 1979, pp. 36-66.

William Vogt, Road to Survival, New York: William Sloane, 1948.

Philip Wagner, The Human Use of the Earth, New York: The Free Press, 1960.

Richard H. Wagner, Environment and Man, New York: Norton, 1974 (1st ed., 1971).

R. O. Whyte and G. V. Jacks, Vanishing Lands, New York: Doubleday, 1939.

Carol Wilson, ed., Energy: Global Prospects 1985-2000, New York: McGraw-Hill, 1977.

W. S. and E. S. Woytinsky, World Population and Production, New York: Twentieth Century Fund, 1953.

Mark W. Zemansky, Heat and Thermodynamics, 5th ed., New York: McGraw-Hill, 1968.

Bruce Hannon

3. The Energy Cost of Energy

In the modern economy, we seem to employ technological strategies to capture, concentrate, and release energy in order to accomplish certain predetermined objectives. Here, these strategies are called processes of energy transformation. Through them, a flow or stock of energy is converted into a form viewed as more useful to the system of which the transformation process is a part. To carry out the transformation of an energy input into a more desirable output (e.g., coal into space heat, mechanical motion, electricity, process heat, and so on), energy flows are required with which to construct facilities and to operate and maintain the transformation processes. These flows are called processing energies.

In theory, it is possible to trace down and add up all the processing energies required from the entire economy needed for the creation and operation of a particular transformation process.* It is also possible theoretically to construct the most efficient alternative within existing technology to convert these processing energies into the desired output. To the extent that the transformation process under examination can produce more of the desired output than that available by using the processing energies in the

*An industrial society also consumes available energy when it extracts minerals from their naturally occurring concentrations. For the U.S. economy in 1967, this quantity was less than 1.5 percent of the fossil fuel consumption [1]. Consequently, such energy forms are not included here. Two assumptions were also made: (1) that the caloric value of the fossil fuels is equal to total availability; (2) that nuclear-produced electricity and hydro-electricity can be counted as though produced by plants using fossil fuels.

least-energy-cost alternate technology, a surplus of the desired output would come into being. Society values any energy-transformation process in proportion to the amount of this physical surplus it can produce. Moreover, the value of the transforming process to society also depends on whether the ratio of the desired output to the value of the most efficient conversion process is rising, steady, or falling over the long run.

Does the transformation process produce a surplus of the desired output? And is this surplus rising or falling with time per unit of input? These questions are the focus of this paper. The first question establishes the feasibility of using a given transformation process; the second question, the desirability of its use. Together, they form a set of necessary conditions for the development of a given energy-transformation process.

This view was common among French physiocrats, such as Cantillon and Turgot, in the middle of the Eighteenth Century. Single-factor theories of value may not explain rational economic behavior today. However, when scarcity problems arise on general inputs that are critical, single factor theories may provide useful insights. In a sense, the physically based view of value and scarcity followed here is akin to the views of Malthus and Ricardo who forecast a diminishing return from land with increasing inputs of capital and labor.

The measure proposed in this paper is similar to the empirical economic measure used by Barnett and Morse, except that they employ cost per unit of extracted resource [2]. The proposal given here is reminiscent of the surprisingly accurate physical-scarcity measure of Hubbert [3] and of the more recent theoretical measures proposed by Fisher [4].

In fact, most of the modern economists try to determine scarcity according to variations in the total extraction or discovery cost for a unit of resource. The procedure is difficult, though, because it is hard to define the stage of resource development that is most appropriate. For example, Kakela [5] has shown that while lean taconite (iron) ore is more expensive than natural ore per unit of iron content, iron produced from the taconite is less expensive in the end. This overall improvement in resource efficiency stems from the fact that the taconite pellets allow a uniform and, consequently, rapid decomposition in the blast furnace. Therefore the best approach is to clearly identify the output of an energy-transformation process and to evaluate changes in the total processing inputs per unit of output. Accordingly, the input fuel being converted by the transformation process directly into the output is not included in the present analysis. This input fuel is seen as flowing from the resource base into the economic system, and the processing energies are thought of as having already been committed to use. The processing energies are seen as originating from within the economic system and as carrying with them the option of being used in alternate devices to produce the desired output. They represent an energy "surplus" formed in a previous period and ready for "investment." The question is where should this potential energy investment go in order to achieve a maximum return.

In this paper, a method is proposed and evaluated by which the feasibility and desirability of energy-transformation processes can be determined. The degree to which a resource is exploited by a particular transformation process is tempered by its total demand for labor (particularly if it is scarce), capital and critical materials, and the environmental impact created by its use.

To avoid dollar measures of transformation processes, a physical measure is employed, in the spirit of Cantillon, Malthus, and Hubbert. The monetary value of energy may not always represent its true value to the society because of subsidies [6], inaccurate pricing techniques and policies, and consumer confusion caused by inflation. For example, new supplies of energy seem to compete on the basis of the marginal dollar cost of energy, while energy-conservation measures tend to compete with the average cost, which generally is lower. This phenomenon occurs because utilities are facing a higher marginal cost for fuel than their customers are. Therefore, saving a unit of energy is a more stringent test of cost-effectiveness than producing a new unit. For system feasibility and desirability, the requisite criterion is one under which a process that saves a unit of energy becomes comparable to processes through which a unit is withdrawn from a stock or captured from a flow of sunlight or from geothermal flows.

Finally, since the dollar cost of energy is not yet a major factor affecting optimum production in cost-minimizing industries, optimal solutions for the minimal use of physical energy are appropriate. For example, Pilati [7] has shown that within 1 percent of the minimum dollar-cost solution for the U.S. paper industry, the physical-energy cost could vary by a factor of 2.5 for the same production level. Therefore, the criterion used here must measure energy in physical units.

This paper probably will be most useful to persons involved in economic planning. The process described may be the most appropriate one for an economy facing the scarcity of a factor that is critical in production. Attempts to estimate the future prices of an increasingly scarce resource proved to be impractical during the 1970's. The strategy proposed here may be the alternative.

Definitions and Assumptions

An energy-transforming system is defined as a process in which an input of a stock or flow of energy is changed into an output of greater utility to the society. "Utility" here has a physical definition: outputs of equal utility have the same thermodynamic availability and the same convenience level -- e.g., a stored or flowing energy, a solid or a liquid fuel, with or without pollution control. The system receives inputs of a raw or semiprocessed form of an energy stock or flow which it will transform directly into an output of higher utility. The system also receives inputs of processing energies from either stocks or flows of energy.

To be complete, this processing must include all the energy needed both directly and indirectly to construct, operate, and maintain the energy-transformation process. For example, the processing energy must include that needed to preprocess and deliver the fuel to the plant or facility in which the energy transformation takes place. The processing energy must include such quantities as the energy needed to mine the iron ore to provide the steel for the construction and maintenance of the transforming facility or complex.

To obtain appropriate estimates for these energy inputs requires using the results of the energy input-output model developed at the University of Illinois [8] or the equivalent. The processing units of the UI model are called primary energy -- meaning that all uses of, say, electricity have been transformed into all the energy that was removed from the ground to produce the electricity.

The processing energies do not include, however, the energy going directly from the energy sectors of the economy to final demand. Therefore, such energy demands as the fuel for work-related travel, exports of coal, and the space heating of government office buildings are not included because of historic national accounting-base conventions. All such energy together with its processing energy amounted to 46 percent of total U.S. energy use in 1967 [8].

Costanza [9] recast the accounting procedure to include this significant amount of energy into the appropriate production processes throughout the economy. He also included an estimate of the solar energy consumed through agricultural processes. Such solar energy and labor-and government-related energy costs are not included in the calculations given here because most of the extant calculations were completed before Costanza's work was published.

Finally, the system might receive a direct input of its own output. This type of input is considered internal to processes of transformation and is reflected in a diminished net output.

The system of energy transformation is shown schematically in Figure 1.

Methodology

The Return on Net Energy Investment

To describe the feasibility criterion, a term, R, is used. "R" is the return on net energy invested: the net output divided by the processing inputs. There are at least three reasonable ways to calculate the ratio. When the ratio is formed for a series of comparable energy-transformation processes, each calculation scheme gives a different ranking of the ratios. Therefore, justification must be given for the scheme selected.

One of the alternate methods is similar to the familiar return-on-investment criterion used by many corporations to rank the potential of investment projects using profits. But the view taken in this paper is that of "society-as-corporation." Therefore, the return should be based on the available surplus, or profit, which society has generated. Under this view, the output of the energy-transformation process must be divided by the processing energies required.

The second alternate method involves consideration of the direct use that a transformation process makes of its own output. This direct input (see Figure 1) could be placed in the denominator of the return ratio, or in the numerator. The latter was chosen here because to group this direct input with the processing energies would be to consider it as a surplus over which society has a choice about use. But it is not such a surplus. This direct input is a function of the transformation process used. The only requirements are that a particular transformation process be used and that its quantity be generally proportional to the output of the process. Thus, this energy is deducted from the gross output of the process and the net output is divided by the processing energies to form a ratio called the "return on net energy investment."

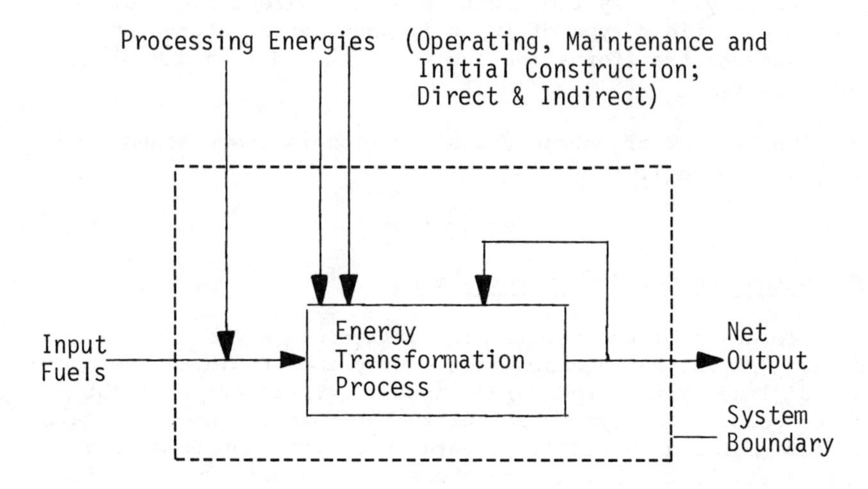

Figure 1. Basic Energy System.

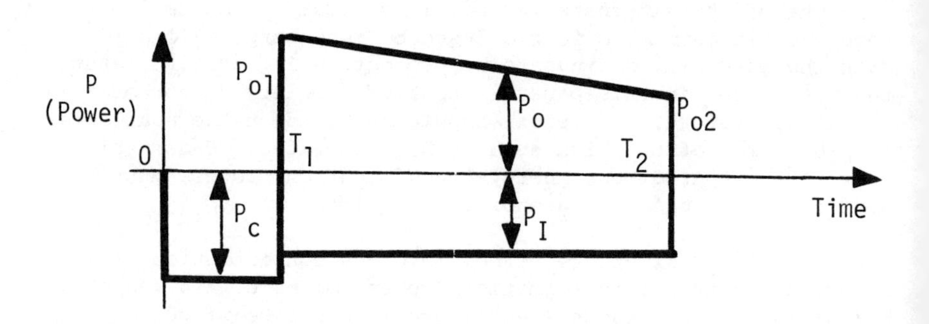

Figure 2. Simplified power curve of a general energy-transformation process in undiscounted units.

For an appropriate ratio, compensations must be made in the numerator and denominator for differences in quality and types of end-use and time of use. To correct the ratio for quality and utility differences, a symbol for total energy efficiency, α, was used. Alpha represents the most efficient means existing of converting primary energy into the exact form as needed as an output.

To correct for differences in the time at which processing energies are committed for use, a standard, continuous-discounting function, $e^{-\lambda t}$, was employed. The discount rate, λ, is the mechanism by which I believe society implicitly expresses its desire to convert a present surplus energy into an energy-transformation process so that a greater surplus of energy can be created in the future, rather than consuming the energy now for purposes such as home heating, leisure living, and certain types of food consumption.

Because of the time element, it is far easier to express the energy uses as a time rate or as power. Figure 2 depicts the highly simplified power diagram of a general energy-transformation process. Figure 2 displays a linearly declining output of a transformation process for constant construction, maintenance, and operating processing energies. The declining output represents an increasingly dilute fuel source, or a declining energy efficiency for an output device. An example of the latter is the tendency of photovoltaic cells to become inoperable after a time [10]. The remainder of the power curve is simplified because available data on energy-transformation processes are not detailed enough in most cases to allow a more elaborate representation.

In Figure 2, P_{o1}, P_{o2} are the initial and final net power outputs at times T_1 (the construction period) and T_2, respectively ($\Delta T = T_2 - T_1$, the lifetime); P_c, P_I represent the primary energy input rates for construction and maintenance and for operation, respectively; m is the slope of the net output rate;

$$m = \frac{P_{o2} - P_{o1}}{T_2 - T_1} .$$

The equation for R, the return on net energy investment, based on the diagram in Figure 2 is:

$$R = \frac{\int_{T_1}^{T_2} P_o \quad e^{-\lambda t} dt}{\frac{P_c}{\alpha}\int_0^{T_1} e^{-\lambda t} dt + \frac{P_I}{\alpha}\int_{T_1}^{T_2} e^{-\lambda t} dt} . \qquad (1)$$

Equation 1 is integrated to give

$$R = \frac{\left[\alpha \ (m\Delta T + \frac{m}{\lambda} + P_{ol}) e^{-\lambda\Delta T} - P_{ol} - \frac{m}{\lambda}\right]}{P_c (1 - e^{\lambda T_1}) + P_I \ (e^{-\lambda\Delta T} - 1)} . \qquad (2)$$

The term α is related to the concept of opportunity cost in economics. The difference between $1/\alpha$ times the present value of the processing powers and the present value of the output power is the opportunity cost of the least-cost alternative in relation to the energy-transformation process under consideration. Through α, the concept of technological substitution is also introduced. The least-cost alternative represented by α is the best substitute process for the proposed energy transformation.

The normalization of the output power by the processing powers in Equations 1 and 2 is similar to the power-gain calculation used by electrical engineers for amplifying circuits. The processing powers are similar to the control power used in such circuits. The engineers are interested in how the output changes with respect to changes in the control power.

The discount rate is, of course, zero in amplifying circuits. A zero discount rate in Equation 1 produces the standard form of the net energy equation commonly used by energy analysts. Most calculations of energy feasibility [10,11] compare the average annual net output with processing input plus the initial construction energy. The construction energy is divided by the expected lifetime of the system. In this scheme, the sequencing of the energy expenditures has no effect on the ratio of net output to net input.

The Feasibility of Energy-Transformation Processes

Energy analysis provides a more-unusual situation than is generally found in economic analyses of benefit and cost. In economics, the discount rate is known. Projects are ranked on the basis of highest to lowest R. In the case of energy discounting, one must determine first the appropriate λ and then use Equation 2 to rank the resulting values of R. To determine the appropriate λ, R = 1 was set in Equation 2. The equation was then applied to all existing energy-transforming processes and solved for λ in each instance.

The smallest λ is assumed to be the energy discount rate presently used by society. This minimum "revealed" λ is used in Equation 2 for all existing and proposed energy-transformation processes to produce a ranked list of R's. Under the definitions used with the theory, society would rather consume a unit of energy now than invest in a transformation process with an R of less than 1. Even though such transformation processes may have a positive R, the amount of surplus energy produced beyond the breakeven point is not large enough to justify the use of that process.

An interesting problem arises if we put the analysis on an identical basis with ordinary financial discounting. Then, the P_I of Equation 1 would be augmented by the rate at which the energy content of the fuel is processed into the output energy. With certain values of P_O, P_C, and ΔT, λ can then become complex or negative, and apparently meaningless.*

Another interesting problem arises that is not pursued here. Equation 2 may be used to determine the maximum R when the size or size-distribution parameters are varied. These parameters are fixed in the examples used in this paper. The question is, for example, would a few, coal-fired electric plants have a higher return on energy investment than many smaller, cogenerating, coal-gas-electric ones when producing the same end-utility?

The value for α must be chosen to represent the most efficient practical way of converting from primary energy to the desired output form. The assumption made is that α applies in Equation 2 when: (1) the output energy is in the form of electricity, $\alpha = 4.0$ [8]; or (2) the output is space heating, $\alpha = 1.5$ [12]; or steam is being produced, $\alpha = 1.15$ [13]; additionally that $\alpha = 1$ in all other processes.

The last value ($\alpha = 1$ in all other processes) is tantamount to assuming that oil, coal, gas, ethanol, wood chips, and so on are of equal utility. The correction is made by calculating the best conversion efficiencies that can be

---

*For a positive and definite λ, $\alpha P_O > P_I \geq 0$ is the necessary condition.

achieved between the input forms and the output form. For the purposes of this paper, the assumption that $\alpha = 1$ in all transformations producing alternate fuel forms is sufficiently accurate.

The efficiency value for electricity production ($\alpha = 4$) reflects the aggregate efficiency of all electric plants using fossil fuels in 1967. As such, the value includes oil- and gas-fired plants which did not need significant pollution-control devices. Many of the coal-fired plants already had precipitators by that time. Therefore, it is assumed here that the value of $\alpha = 4$ is sufficiently representative of the least-energy cost alternative to all of the options for producing electricity.

In converting the process inputs into the desired type of output, the efficiency $1/\alpha$ includes the energy costs of the capital needed directly and indirectly for the conversion process. This capital energy cost is annualized and is added to the operating and maintenance energy to produce α. Although the effect of time should be considered in the α calculation, that effect is omitted here because of its small contribution to R in Equation 2. Including capital energy effects influences the energy costs of the average product produced by about 7 to 8 percent [14]. Here α includes the capital effects, but in an approximate way ($\lambda = 0$). One can arbitrarily limit the error by an iteration of the trial-and-error solution of Equation 2.

The value of α could be reduced to represent the energy lost by placing and maintaining the output in storage. Conversely, the amount of fossil fuel backup needed to provide the desired output pattern could be added to P_I. Either way, the return from the net energy investment is reduced.

The Effects of Scarcity on the Desirability of an Energy-Transformation Process

The foregoing analysis allows us to determine the feasibility of a particular transformation process at a given state of energy-resource availability. The analysis also allows us to rank those transformation processes that produce a certain desired type of output (e.g., electricity, by a descending order of feasibility.

If net energy efficiency were the only consideration, such a ranking would show the order to be followed in the development of energy-transformation facilities, that is, the relative desirability of each process. However, as each transformation process deemed thus to be feasible were to operate on the resource base, its R value eventually would decline while the facility would continue to produce a constant output.

The decline may occur because λ would be increasing (a change in social objectives, perhaps driven by a declining concern about scarcity or political instability of the input fuel supply), or because α would be decreasing (improvements made in the energetic efficiency of the least-energy-cost alternative), or perhaps because the amount of processing power per unit of output power would be increasing in a present-value sense (input scarcity). These three factors embody changes in consumer behavior, transforming technology and energy scarcity.

The processes of energy transformation with the highest, most slowly declining R values are the most desirable ones. Among these technologies, those that depend on politically stable supplies of fuels and materials and those which make socially agreeable demands on the labor force, capital market and the environment are the ones that should be developed intensively. Transformation processes with R values that are close to or less than 1 should be dropped from consideration and be the object of further research and development.

Even so, we still do not know how to phase together the ranked technologies for energy transformation so that the net energy with the greatest present value is produced. To accomplish this phasing, assume that the only reason for the decline in R values is a scarcity of input energy, for example, the coal available to a power plant in a certain region might be found to be in thinner and deeper seams. Assume also that the peak efficiency of scale in energy production had been reached, for example, an incremental increase in the energy efficiency of a large power plant in a certain region would be exactly offset by a decline in the energy efficiency of coal-gathering and electricity transmission. Let the cumulative present value of the output be f and the cumulative present value of the capital, operating and maintenance processing powers (energy) be g. Both f and g are energy measures. Each transformer has a characteristic relationship, as shown in Figure 3, because of the two assumptions just made. The three curves in Figure 3 could represent, for instance, the use of oil from Texas, the Middle-East, and Alaska to move a ton-mile of rail freight with a diesel-electric engine that burns oil. As is easily proven, the calculation for R is independent of the time lapse from $f=g=0$ (R depends only on time periods T_1 and ΔT). Therefore, the curves $f = f(g)$ do not depend on the time lapse. Consequently, the curves in Figure 3 can be constructed from the present value

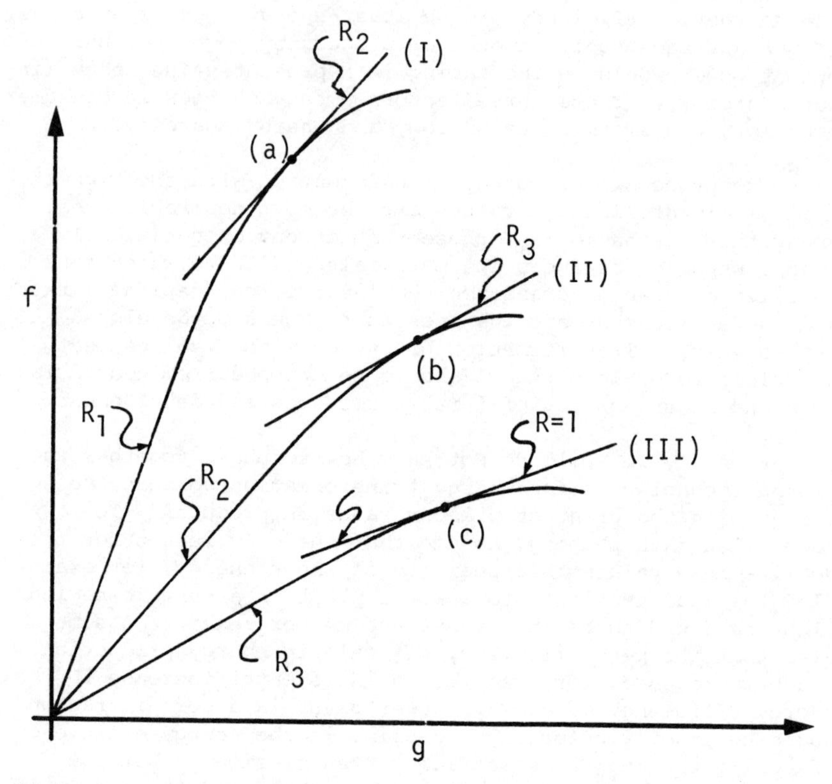

Figure 3. The cumulative present-valued output energy, f, vs. the cumulative present-valued processing energy, g, for three energy-transforming technologies.

of f and g calculated for each energy-transformation process independent of its time sequencing.

The derivative df/dg at any point along the curves for any of the three transforming technologies is the return on net energy investment, R. The process labeled (I) would be the logical place to begin for it has the highest initial value of R ($R=R_1$) that is possible for all three processes. However, when the cumulative energy output reaches point a, it would be appropriate to switch to the second transforming technology (II). At point a, the initial R of the second transforming process would be equal to the current R ($R=R_2$) for the first process. The second process would be employed until its R value declined to point b, where it becomes equal to the highest ($R=R_3$) value for the third process (III). This pattern would be continued through the range of available transforming technologies until the present-valued R reached 1 (point c).

If the discounting value or the least-cost-alternative efficiency should change, Figure 3 would have to be reconstructed. The above procedure would still be followed, but using the newly computed information.

Results

Data for the feasibility calculation only were found on 44 energy-transforming processes (Table 1). Only the solar-powered satellite transformation process showed a decline in output power with constant energy for maintenance and operation. This decline is not an example of the scarcity phenomena already described. Rather, it is typical of an individual satellite unit. Only if the allotted section of the geosynchronous orbit were filled with such units would scarcity begin to lower the R values. None of the examples identified in this study showed scarcity responses.

Most of the examples were based on the University of Illinois Energy Input-Output model [8], but only one contained an indication of error tolerances [10]. However, the individual authors used different techniques to assign joint energy costs and to credit byproducts. Every possible effort was made to place the data in a common framework. Yet these problems and the assumptions stated earlier make the results in Table 1 an interesting experiment at best.

No firm policy conclusions should be inferred without further research. For the purposes of demonstrating the potential usefulness of this approach, however, the results were analyzed as though the data were sufficiently accurate.

Table 1. Calculation of the Return on Net Energy Investment, R (from Equation 2).

Process	YEARS		α	P_o	P_1	$P_c \cdot T_1$ (a)	$\lambda(\frac{1}{Yr})$ Percent (R=1)	R (λ=0.19)	R λ=0	COMMENT [Reference]
	T_1	ΔT								
FUEL PRODUCTION										
-Nonrenewable										
Methanol from Natural Gas	3	20	1	1	.42	.20	76	2.1	2.3	Not used as a fuel [15]
Natural Gas	5	30	1	1	.17	1.32	28	1.7	4.7	Includes delivery [15]
Gasoline	4	25	1	1	.074	.63	48	3.9	10.1	Includes delivery [15]
-Renewable										
Intensive Tree Plantation	.5	15	1	1	.20	.091	340	4.6	4.9	Jack Pine; Chipped, Dried [17]
Natural Tree Plantation	0	15	1	1	.21	0	...	4.8	4.8	Chipped, Dried [18]
Crop Plantation (for fuel):										
Corn	.5	10	1	1	.35	.013	660	2.8	2.9	New Mexico, Fert. & Dry [19]
Wheat (Dry)	.5	10	1	1	.19	.067	390	4.9	5.1	New Mexico, Fert. & Dry [19]
Wheat (Irrigate)	.5	10	1	1	.48	.043	390	2.0	2.1	New Mexico, Fert. & Dry [19]
Sorghum (Dry)	.5	10	1	1	.25	.043	455	3.8	3.9	New Mexico, Fert. & Dry [19]
Sorghum (Irrigate)	.5	10	1	1	.31	.031	500	3.2	3.2	New Mexico, Fert. & Dry [19]
Alfalfa	.5	10	1	1	.37	.051	394	2.6	2.7	New Mexico, Fert. & Dry [19]
Peanuts	.5	10	1	1	.89	.08	105	1.1	1.1	New Mexico, Fert. & Dry [19]
Methanol (Wood)	4	20	1	1	.34	1.00	32	1.6	2.6	Coal-fired distillation. Low R does not include Octane Credit [15]
Ethanol (Corn)	4	20	1	1	.75	.75	21	1.03	1.3	Low R, does not include Octane Credit [20]

(a) To put this column into average power terms, divide by T_1.

Table 1, continued

Process	YEARS		α	P_o	P_1	$P_c \cdot T_1$ (a)	$\lambda(\frac{1}{Yr})$ Percent (R=1)	R (λ=0.19)	R λ=0	COMMENT [Reference]
	T_1	ΔT								
ELECTRICITY PRODUCTION & DISTRIBUTION										
-Nonrenewable										
Coal	5	25	4	1	0.09	0.48	75	16.4	36.6	U.S. Average [11] } No scrubbers
Combined Cycle (Coal)	4	25	4	1	0.03	0.84	75	14.8	62.9	[11] } No scrubbers
Eastern Surface Coal (b)										
No Scrubber	5	30	4	1	0.06	1.00	61	10.6	42.9	} 300 mile coal haul [21]
Limestone Scrubber (c)	5	30	4	1	0.34	1.10	57	5.8	10.6	} 300 mile coal haul [21]
Western Surface Coal (b)										
No Scrubber	5	30	4	1	0.13	1.11	58	8.3	24.0	} 1120 mile coal haul; West. coal not as intensively scrubbed as East. coal to meet same emission standards [21]
Limestone Scrubber (c)	5	30	4	1	0.36	1.20	56	5.4	10.0	}
Coal; w/Limestone Scrubber (e)	5.5	30	4	1	0.19	0.80	60	8.7	18.5	} [22] Less complete analysis than [21].
Coal: Fluidized Bed (Pressurized) (e)	5.5	30	4	1	0.12	0.82	60	10.1	27.1	}
Solvent Refined Coal	5	25	4	1	0.38	0.72	66	6.6	9.8	[11]
Coal-Gas	5	25	4	1	0.23	0.96	61	7.4	14.9	[11]
Shale-Gas	5	25	4	1	0.52	0.84	62	5.1	7.2	[11]
Oil-Gas	5	25	4	1	0.52	2.04	45	3.4	6.7	[11]
Natural Gas	5	25	4	1	0.32	1.08	58	6.0	11.0	[11]
Nuclear										
LWR (d)	6	25	4	1	0.14	11.27	19	1.0	6.8	[11]
HTGR (d)	6	25	4	1	0.11	18.30	13	0.6	4.8	[11]
Geothermal (e)	3.1	30	4	1	0.18	2.28	59	5.2	15.6	Liquid Dominated, High Cost Option [23]

(a) To put this column into average power terms, divide by T_1.
(b) Surface mine R values only 2% higher than underground R values. A variation in the construction period T_1 of ± one year gave a variation in R of from ± 6% to ± 10%, the higher variation being associated with the higher R values. (e)
(c) A Wellman-Lord scrubbing process lowered R by 28% while the limestone scrubbers lower R by 43%. (e)
(d) If the construction period were 9 years instead of 6, the resulting λ would be 16% and 12%, respectively.
(e) The input requirements of these processes were increased by 9% to reflect the losses in transformation and transmission [27]. Data from Reference [11] contained this correction.

Table 1, continued

Process	YEARS T_1	YEARS ΔT	α	P_o	P_I	$P_c \cdot T_1$ (a)	$\lambda(\frac{1}{Yr})$ Percent (R=1)	R (λ=0.19)	R λ=0	COMMENT [Reference]
ELECTRICITY PRODUCTION & DISTRIBUTION										
-Renewable										
Wood (No pollution control)(e)	5	30	4	1	0.802	0.59	67	4.0	4.9	Assumed eff. = 0.3 [24]
Solar Power Satellite(e)	5	30	4	1.00 / 0.44	0.008	9.9	21	1.1	8.5	Output decline [10]
Photovoltaic (No storage or backup fuel or transmission costs)	0.2	20	4	1	0	12.0	32	1.7	6.7	[25]
Wind (No storage or backup fuel or transmission costs)	0.4	30	4	1	0	0.15	614	135.0	800.0	Est. Construction Time [26]
Power Tower (No storage or backup fuel or transmission costs)	5	30	1.15	1	0	0.54	49	6.7	64	Test facility; steam production only [25]
SPACE HEATING										
-Nonrenewable										
Gas Furnace	0.25	20	1.5	1	0.15	0.22	38	7.7	9.3	[28]; Operation Energy [8]
Electric										
Nuclear LWR	6	25	1.5	1	.21	10.53	8	0.4	2.4	
Coal	6	30	1.5	1	0.307	1.00	35	2.3	4.4	Eastern surface mine, conventional scrubber[21]
-Renewable										
Solar										
Flat Plate Collector	.5	20	1.5	1	0.56	5.05	17	0.9	1.9	Fossil fuel backup [28]
	.5	20	1.5	1	0.60	6.89	11	0.7	1.6	Fossil fuel backup [29]
Concentrating Collector	.5	20	1.5	1	0.56	4.31	20	1.0	1.9	Fossil fuel backup [28]
Direct Solar Gain	.5	20	1.5	1	0.4	0.93	93	2.5	3.4	Fossil fuel backup [28]
CONSERVATION										
Double Pane Windows	.1	30	1	1	0	0.22	375	24.0	136.0	Chicago [30]
Ceiling Insulation (Retrofit-National Program)	7	30	1	1	0	0.49	39	5.1	61.0	Retrofit National Program [31]
Urban Car to Bus	1	14.8	1	1	0	0.44	118	10.2	34.0	[32]
National Energy Plan I	5	20	1	1	0	0.12	76	26.4	171.0	Very approximate [33]

(a) To put this column into average power terms, divide by T_1.

(e) The input requirements of these processes were increased by 9% to reflect the losses in transformation and transmission [27]. Data from Reference [11] contained this correction.

The examples in Table 1 were divided according to the output of the process: fuel production, electricity production, space heating, and conservation. The first three are segmented into nonrenewable and renewable sources of fuel.

Even though a certain process may have a very high return on the energy investment, the potential for that process in terms of producing significant additional quantities of useful energy may be limited. For example, in the Unites States, most large and convenient sites for the production of hydroelectricity are occupied; and yet, because of the negligible operating and maintenance energy, hydropower could have a very high R value. Additional units of such processes should show a rapidly declining R (calculated) as a result of increases in the capital requirements and energy-transmission losses per unit of output.

Choosing the Discount Rate

Equation 2 with R value equal to 1 was applied to each process and solved for λ by trial and error.* The low end of the resulting λ spectrum contains both solar and nuclear processes. The lowest λ value (λ = 13 percent) among the electrical generators, is for the high-temperature, gas (nuclear) reactor (HTGR). This process, like the light-water (nuclear) reactor (λ = 19 percent) has a low value for λ because of the large capital outlay required before operation begins. The HTGR process is not in general use today in the U.S. so it is dropped from further consideration.

Note that if electricity produced by the light-water reactor process is used for resistance heating, the resulting λ is about 8 percent.** If a particular end use of a technology is not known by society, then that end use could not enter society's choice of a lowest acceptable discount rate. Since it is not clear that society is aware that LWR electricity is being used for resistance heating (because of the mix of plant types on the same electricity-distribution system), the higher value of λ was chosen for the light-water reactor as the base for all calculations on energy returns.†

---

*Remember that none of the processes considers the energy content of the fuel which is processed into the final energy output.

**From equation 2, we see that lowering α by 63 percent (4 to 1.5) drops the R value by 63 percent.

†If electricity were not subsidized [6] and were priced marginally rather than on an average [34], only minor amounts of electricity might be used for heat.

The flat-plate solar collectors (λ = 11 and 17 percent) and the solar-powered satellite (λ = 21 percent) have low λ values because of the large initial capital investment. The solar powered satellite also has a declining output over its lifetime because of the slow, irreversible degradation of the photovoltaic cells caused by cosmic radiation. The discounting process makes this decline less important than it would be if λ were zero.

The solar-concentrating collector (λ = 20 percent) provides only a small improvement over the flat-plate collector. Ethanol from corn (λ = 21 percent) owes its low λ value to the high operational energy generally employed to fertilize the corn and to distill the alcohol.* Methanol from wood (λ = 32 percent) is a superior process to ethanol (corn) in overall energy efficiency and has the same λ as earth-based photovoltaic cells. Natural-gas production (λ = 28 percent) is a nonrenewable energy source with a surprisingly low λ value. The relatively high λ values for photovoltaics, windmills, and the solar power-tower would be reduced considerably if energy costs for storage were included.

Therefore, the only appropriate solution to Equation 2 with R = 1 was for the light-water nuclear reactor. This is the only technology with a low λ value that is presently used to transform significant quantities of primary energy into desired forms of output (electricity). The discount rate for this technology (λ = 0.19) represents the lowest that society will accept, assuming that such a rate even exists. No other widely used technology has a lower λ value.

How sensitive is this rate to changes in the key variables? We can define a standard elasticity equation as:

$$e_{T_1\lambda} = \frac{T_1}{\lambda} \frac{\partial\lambda}{\partial T_1}, \qquad (3)$$

which is the elasticity of λ with respect to T_1. A small change in T_1 will produce an $e_{T_1\lambda}$ change in λ.

---

*Suppose that the distillation process involved in making ethanol from corn had been solar-heated. How would this energy been included in Equation 2? First, the heat energy absorbed would have been included in the operating energy. Second, the α would have to be modified to contain the efficiency of the conversion of that solar radiation absorbed to a similar form of output (possibly dried biomass).

Table 2. Elasticities for the Key Variables in Equation 2 (R = 1) applied to the Light-Water Nuclear Reactor process (λ = 0.19)

T_1	ΔT	P_o	P_I	P_c
-0.995	-0.001	+0.960	-0.960	-0.431

Equation 3 was applied to each of the variables. The results are given in Table 2.

For example, an increase of 1 percent in T_1, the construction period, would reduce λ by 0.995 percent. A 1 percent increase in the energy rate used in the construction period, P_c, would reduce λ by 0.431 percent. The lifetime, ΔT, is not an important variable in solving for λ; but T_1, P_o, and P_I are important.

The values in Table 2 reveal the sensitivity of λ to small changes. For a large change, Equation 2 (R = 1) must be solved again for λ by trial and error. When this was done, the λ sensitivity was not great. For example, a change of 10 percent in each variable produced a maximum change in λ of 2.5 percent. It is safe to conclude that λ is not very sensitive in the variables T_1, ΔT, α, P_o, P_I, or P_c; also, that T_1 is the most important variable with regard to the determination of λ.

The construction period, T_1, is the least accurate of the variables in Table 1. T_1 is derived from pure guesswork in some instances, to averages of actual practice recently. In every case, the construction period is that time during which $P_c T_1$ is expended.

Another assumption is that indirect capitalization takes place during T_1. Here is a key problem in this kind of analysis. The time and energy spent on the direct and indirect capitalization must be assigned to the joint products on what amounts to an arbitrary basis. The longest of these assigned periods is T_1. The shorter periods are superimposed in the most practical possible order. This procedure may not represent reality since iron must be made before steel, which is needed for sheet steel. That, in turn, is necessary for power-plant fabrication, and so forth. The practical time for these processes may actually exceed T_1. However, the construction period encompasses the bulk of the capital-forming time for any given process of energy transformation and is, therefore, sufficiently accurate for the purposes here.

An additional aspect of the concept of energy opportunity cost must also be recognized. As noted earlier in this paper, the choice of the most energy efficient converter for the processing energies does incorporate the idea of energy opportunity costs into the choice of the energy discount rate. But this incorporation has not been proven to be complete. It is possible that the use of the processing fuels might ultimately save more primary energy if it were invested directly in a conservation project (e.g., to harvest more wood to displace the use of energy intensive concrete in building construction). The resulting saved-energy could then be applied to the most efficient converter and compared with the energy output of the transformer in question. There are two aspects of the problem which tend to rule out all such opportunity cost from consideration. First the "construction period" is exceedingly long because the processing energies have to be invested in programs whose payoff is over a lengthy period. Thus rather than an "instantaneous" least cost conversion of these energies as is considered in this paper, the least cost alternative payoff must be discounted. Secondly, the conservation project alternative has an ultimate saturation limit which might be quite small. The least cost alternative process used here converts processing energies into the same form as the process of transformation under consideration. As long as there is a demand for the output of this particular process, there will likewise always be a demand for the output of the least cost alternative process.

Finally, we must realize that the λ discussed in this paper is thought of as an aggregate value for the whole society. Regional variations in the discount rate are entirely possible. These variations might arise from differing local views about the certainty of future fuel supplies and their future costs. The concepts set forth in this paper are so tentative that delving into regional differences in discount rates must wait for appropriate opportunities with suitable data.

The Energy Return on Net Energy Investment

When we compare the R values, what do we find? The liquefied fossil fuels appear to be superior to the liquefied fuels made from vegetables. Among the renewable fuels, trees are better than field crops such as corn, whether one is making a solid or a liquid fuel. This is probably because a young to medium-aged forest can convert more solar radiation into storage per year and per acre than a cash crop can.

Irrigated crops have lower R values than the same crop without irrigation. Here we see the tradeoff between the fossil fuels (irrigation) and land (yield). Finally, on the basis of 1 data point (peanuts), it would seem that the grass crops (corn, wheat, sorghum) are better than legumes. The production and distribution of natural gas has a surprisingly low R value because of the extensive investment required in pipelines and the high energy costs associated with pumping operations.

In the production of electricity, fuels and geothermal sources appear to be superior to nuclear and solar-based processes. The possible exceptions are windmills (no storage costs included) and wood. Coal-fired power plants without environmental-protection equipment are the most feasible processes (e.g., the average conventional coal-electric and combined-cycle processes).

In a more detailed view, we compared coal-electric processes in the Eastern and Western United States, with and without appropriate scrubbers. Scrubbers reduce the effectiveness of coal-electric plants from 50 to 75 percent. This efficiency reduction is one way to assess the energy cost of air-pollution abatement. With scrubbers, Eastern coal is slightly more energy-effective than Western coal. The energy return is not affected by the type of mine.

The pressurized fluidized bed coal-electric process,* prescribed because of its low levels of harmful effluent, has an R value that is 16 percent higher than that of a scrubbed Eastern coal-electric plant. To control coal-associated air pollution, the solvent-refined coal and coal-gas processes are approximately as feasible as using either the scrubbers or the fluidized-bed processes. The shale-gas-electric process is better than the (heavy) oil-gas-electric process.

The natural gas-electric process, normally thought of as attractive because pollution control is achieved without a major investment, has a similar energy return to the scrubbed-coal electric or the fluidized-bed coal processes under the analysis presented here. The geothermal-electric process, which is free of air pollution, is classed as nonrenewable because the resource rock is actually mined of its heat content. This process has a high R value, comparable to the coal-gas-electric process and the solvent refined and scrubbed-coal processes.

---

*This process and its companion in Table 1 were not analyzed in as much detail as the others [10, 11, 23]. They appear only to give the relative energy efficiencies.

Of the renewable electric processes, the photovoltaic, wind, and power-tower do not store energy. Consequently, they cannot be compared to the other process in Table 1. The wind machine is a superior energy-converter compared to the photocell or the concentrated solar-heated boiler (power-tower). The wood-electric and the solar satellite are comparable processes because, theoretically, the electric energy is available on demand from either process. The wood-based process appears to be superior, although reliable data for either process are not available.

The last comparison points up one of the most important technological questions of our time. Can man devise a renewable energy source to meet an arbitrary demand cycle that is superior in energy efficiency to the natural processes of capturing, storing, and releasing high-quality energy? The data in Table 2 indicate that we have not yet managed to duplicate the efficiencies of natural energy. However, the wood-electric process does not include pollution-control effects and is, therefore, not strictly comparable to a solar satellite as an energy source.

None of the solar space-heating options compare well with the gas-furnace heating system. The direct-gain, or passive solar space-heating process is energetically superior to the flat-plate or the concentrating solar collector and is approximately equivalent to the scrubbed coal-electric heating process. All of the space-heating processes investigated had a greater net energy return than the light-water nuclear reactor.

Note that in this paper, the solar processes use gas heaters to provide heat when solar radiation is not available. By this procedure, one of the energy costs of convenience (demands for energy that differ from the natural solar (cycle) can be estimated.

The conservation projects appear to be feasible in terms of overall energy efficiency. They compare well with fossil-based processes in which pollution-control strategies are used. The conservation processes are somewhat difficult to fit into the simplified graph in Figure 1 because they are national programs with effects that are being felt before the construction programs are finished. This was true of the ceiling-insulation program and of National Energy Plan I.

The phenomenon of the extended process points up an interesting and complex problem. Both conservation and supply-construction programs can theoretically proceed at rates that will not produce a net energy output. In general, the data in Table 1 are achieved only when building rates slow toward a replacement level. Clearly, a proper energy analysis would be applied to the entire program from initiation until a steady-state condition is reached -- with the energy costs and benefits being placed in present-value terms.

Also note the effect on R values of different λ values. When $\lambda = 0$, for example, Equation 1 reduces to the form commonly used by energy analysts. The ranking of the R values in Table 1 is changed substantially when $\lambda = 0$, mainly because some processes have long construction periods with heavy capital commitments in relation to others.

If the λ value and the real financial discount rate are equal, then financial feasibility necessarily means energetic feasibility, providing that the economy can be thought of as having a single energy source [35]. However, the physical discount rate now appears to be several times higher than the real financial rate, so we can imply nothing about physical feasibility from financial feasibility. Eventually, it seems possible that the two rates may converge as the stocks of energy resources are reduced.

If the λ value does represent the society's time value of energy, then it should rise during periods when energy is perceived as being plentiful and should decline during periods of energy scarcity or when the input fuel supply seems unstable due to political or social unrest. From the values in Table 1, we find that a society which burns wood has a value for λ of 67; in a coal-burning society, λ ranges from 75 down to 55 with pollution control; and in the nuclear age, we have a λ value of 19. The minimum λ seems to have generally declined in the United States over the past 100 to 125 years. From the consumer's viewpoint, this phenomena would indicate a gradually increasing scarcity of available energy. Eventually, as fossil and nuclear fuels are depleted, the minimum λ will decline until it reaches the level for the steady-state, agrarian society, probably zero. No surplus energy would be needed to replace the depreciating structure and provide for all aspects of the renewal cycle of society and family life. These needs would be considered part of the maintenance energy demand.

Also recall the classic argument of Page [36], who said that owing to the unavoidable decline in the availability of resources because of use by the present generation, the only way in which we could be totally fair to future generations is to leave them the tools and the knowledge to derive as much benefit as we could have done originally, using the known resources that remain. To accomplish this goal, no better procedure than the one given here seems to be conceivable.

Summary

The present value, net energy feasibility of 44 energy-transforming processes was calculated and compared on a purely energy-efficient basis. The method allows a quantitative comparison of all forms of supply and end-use processes -- whether they are from renewable or nonrenewable sources of energy, whether they are fuel, electricity, space heating or transport end-use, or whether the energy gain results from a substitution that frees energy for use elsewhere. Processes using fossil fuels generally are superior to most of the solar and nuclear ones. With certain noted exceptions, the conservation processes studied are comparable to fossil-fuel processes.

The data are of limited validity, not allowing any firm policy conclusions to be suggested. Even so, a useful method of energy analysis, I believe, has been created and demonstrated. The availability of suitable data with the requisite validity would permit the analytical procedure to be applied broadly in the process of formulating and implementing energy policy.

Acknowledgments

I wish to thank Horacio Perez-Blanco for his part in data gathering and interpretation and to thank him, Robert Costanza, and Andrew Postlewaite for their helpful discussions.

References

1. Joyce, J., "Exergy," Tech. Memo 120, Energy Research Group, University of Illinois, Urbana, IL 61801, Jan. 1979. For a contrary view, see the elegant paper: Georgescu-Roegen, N., "Energy Analysis and Economic Valuation," Southern Economic Journal, V. 45, No. 4, April 1979, p. 1023-1058.

2. Barnett, H. and Morse, C., "Scarcity and Growth," The Economies of Natural Resource Availability," Johns Hopkins University Press, Baltimore, 1963. See also: Barnett, H. "Adequacy of Scarcity Measures," Chapter 8 in Scarcity and Growth Reconsidered, K. Smith, Editor, Johns Hopkins University Press, Baltimore, 1979.

3. Hubbert, M. "Degree of Advancement of Petroleum Exploration in the U.S." Am. Assoc. of Petroleum Geologists Bulletin, Vol. 51, No. 11, 1967, pp. 2207-2227. Also see his seminal article in Science, Vol. 109, p. 103, 1949.

4. Fisher, A., "Measures of Natural Resource Scarcity," Chap. 10 in Scarcity and Growth Reconsidered, K. Smith, editor, Johns Hopkins Uniersity Press, Baltimore, MD, 1979.

5. Kakela, P., "Iron Ore: Energy, Labor, and Capital Changes with Technology," Science, Vol. 202, No. 4373, pp. 1151-1157, December 1978.

6. Bezdek, R., and Cone, B., "Federal Incentives for Energy Development," Energy, Vol. 5, May 1980, pp. 389-406.

7. Pilati, D., Letter and data to B. Hannon. Brookhaven National Laboratory, Upton, N.Y., 15 April 1980.

8. Bullard, C.W., Penner, P.S., and Pilati, D.A., "Net Energy Analysis Handbook for Combining Process and Input-Output Analysis," Resources and Energy 1 (1978) pp. 267-313.

9. Costanza, R., "Energy Costs of Goods and Services in 1967 Including Solar Energy Inputs and Labor and Government Service Feedbacks," ERG Doc. 262, Energy Research Group, University of Illinois, Urbana, IL 61801, September 1978.

10. Herendeen, R., Kary, T., and Rebitzer, J., "Energy Analysis of the Solar Power Satellite," Science, V. 205, No. 4405, Aug. 3, 1979, pp. 451-454.

11. Pilati, D., "Energy Analysis of Electricity Supply and Energy Conservation Options," Energy, V. 2, No. 1, pp. 1-7, 1977.

12. O'Neal, D.L., "Energy and Cost Analysis of Residential Heating Systems," Report No. ORNL/CON-25, Oak Ridge National Laboratory, Oak Ridge, TN 37830, July 1978. The average seasonally adjusted furnace efficiency was 0.65.

13. Hannon, B., and Joyce, J., "Energy Conservation Through Industrial Cogeneration," forthcoming in Energy. Copy available from author.

14. Kirkpatrick, K., "Effect of Including Capital Flows on Energy Coefficients, 1963," CAC Tech. Memo. 26, Energy Research Group, University of Illinois, Urbana, IL 61801, August 1974.

15. Hannon, B., and Perez-Blanco, H., "Ethanol and Methanol as Industrial Feedstocks," Report to Argonne National Laboratory, Argonne, IL, Contract No. ANL 31-109-38-5154, September 1979.

16. "Net Energy Analysis, An Energy Balance Study of Fossil Fuel Resources," Colorado Energy Research Institute, Colorado Springs, CO, April 1976.

17. Zavitkovski, J., "Energy Production in Irrigated Intensively Cultured Plantations of Populus, Tristis # 1 and Jack Pine," Forest Science, Vol. 25, No. 3, 1979.

18. Blankenhorn, P., Murphy, W., and Bowerlox, T., "Energy Expended to Obtain Potentially Recoverable Energy from Forests," TAPPI, Forest Biology-Wood Chemistry Conference, Madison, Wisconsin, 1977, p. 277.

19. Patric, N., "Energy Use Patterns for Agricultural Production in New Mexico," Agriculture and Energy, Academic Press, 1979, Chapter 5.

20. Chambers, R.S., Herendeen, R.A., Joyce, J.J., and Penner, P.S., "Gasohol: Does It or Doesn't It Produce Positive Net Energy?" Science, V. 206, 16 November 1979, pp. 789-795.

21. Penner, P., Kurish, J., and Hannon, B. "Energy and Labor Cost of Coal Electric Fuel Cycles" ERG Doc. 273, Energy Research Group, University of Illinois, Urbana, IL 61801, April 1980.

22. Whittle, C., and Cameron A., "Energy Requirements for Fluidized Bed Coal Combustion in 800-1000 MW Steam Electric Power Plants," Institute for Energy Analysis, Oak Ridge, TN., February 1977.

23. Herendeen, R. and Plant, R., "Energy Analysis of Geothermal Electric Systems," ERG Doc. 272, Energy Research Group, University of Illinois, Urbana, IL 61801, December 1979.

24. Gunn, T.L., "The Energy Optimal Use of Waste Paper," ERG Doc. 263, Energy Research Group, University of Illinois, Urbana, IL 61801, November 1978.

25. Grimmer, D., "Solar Energy Breeders," Los Alamos Scientific Lab., Los Alamos, NM, Report No. LA-UR-78-2973, 1978.

26. Devine, W., "An Energy Analysis of a Wind Energy Conversion System for Fuel Displacement," Institute for Energy Analysis, Oak Ridge, TN, February 1977.

27. Edison Electric Institute, "Statistical Yearbook for 1976," Washington, D.C., 1977. Tables 10s and 20s. Average for 1975 and 1976.

28. Sherwood, L., "Total Energy Use of Home Heating Systems," Symposium on Energy Modeling and Net Energy Analysis, Colorado Springs, CO, August 21-25, 1978.

29. Rogers, D., "Energy Resource Requirements of a Solar Heating System," National Research Council of Canada, Report, Ottawa, February 1979.

30. "Handbook of Energy Conservation for Mechanical Systems and Buildings," R. W. Roose, ed., Van Nostrand, New York, 1978, Chapter 9.

31. Ford, C., and Hannon, B., "Labor and Net Energy Effects of a National Ceiling Insulation Program," to appear in Energy Systems and Policy, Energy Research Group, University of Illinois, Urbana, IL 61801, September 1978.

32. Hannon, B., and Puleo, F., "Transferring from Urban Cars to Buses: The Energy and Employment Impacts," The Energy Conservation Papers, (Robert H. Williams, ed.) Ballinger Publishing Company, Cambridge, MA, 1975, Chapter 3. (A report to the Energy Policy Project of the Ford Foundation.)

33. Hirst, E., and Hannon, B., "Effects of Energy Conservation in Residential and Commercial Buildings," Science, V. 205, August 17, 1979, pp. 656-661.

34. Moden, B., Hatcher, D., and Walton, H., "Marginal Costs of Energy in 1979," Dept. of Energy, Washington, D.C., DOE/EIA-0184/18, October 1979.

35. Bullard, C., "Energy Cost and Benefits; Net Energy," Energy Systems and Policy, Vol. 1, No. 4, 1976.

36. Page, T., "Conservation and Economic Efficiency, Johns Hopkins University Press, Baltimore, MC, 1977.

Kenneth J. Arrow

4. The Response of Orthodox Economics

Once again, we have much to learn from the original and penetrating contributions of Nicholas Georgescu-Roegen. Earlier he had brought the attention of economists to the importance of the entropy concept as the physical counterpart of the economist's concept of scarcity. In the paper in this volume and in earlier papers (1), he has brought these concepts and others to bear on the economics of energy. The full development of his models brings out many points, some of which I shall refer to later, but the two leading implications for understanding the economics of energy are (1) a protest against the "price complex" as a full guide to decision-making about energy, and (2) the proposition that any economic system is inevitably a consumer of matter in useful form. In effect, these two statements stake out a middle ground between two contending schools of thought. The neoclassical or "orthodox" position (I play the role of advocate of this view) is that there is no special economics of energy, that the rule of maximizing profits or minimizing costs at market prices provides an efficient allocation of resources. The "energy" position is that every good should be though of as measured by the energy directly or indirectly contained in it, and the aim of policy should be to minimize the energy content of goods produced. The attack on the "price complex" is a criticism of the former view, the emphasis on ultimate need for matter contradicts the view that only energy counts.

Bruce Hannon's paper is precisely a calculation of energy content. It does make one important contribution: it emphasizes the time dimension. Inputs of energy frequently precede outputs, at least in part; thus in producing electric energy by any of the conventional means, there is a period in which energy is expended for construction, before any energy is obtained. Further energy inputs are simultaneous with outputs. Since both input and output flows are streams over time, he commensurates them by devices familiar from the theory of capital. Thus, the internal rate of return (not his term) can be calculated as that rate which makes the sum of discounted inputs equal to the sum of discounted outputs. The internal rate of return on a project that is actually used to some significant extent can be regarded as an accepted rate, analogous to the market rate of return; then, another calculation can be made which discounts both outputs and inputs at this accepted rate and then takes the ratio of discounted output to discounted input. This ratio is formally analogous to the benefit-cost ratio supposed to be used for project acceptance in the public sector, but it takes account only of energy inputs and those defined in a very particular way.

What is the purpose of these calculations? Hannon is not very explicit but does suggest that as energy becomes more costly relative to other factors of production, net energy becomes an increasingly more important criterion for decisions about energy supply systems. He recognizes that energy has a quality or utility dimension, and therefore it is useful energy that is considered. He appears to suggest that decisions about future energy sources should be guided by the calculated ratios of discounted energy output to discounted energy input.

But there are clear difficulties with using the calculations for these purposes. Georgescu-Roegen's work shows that matter is inevitably degraded; surely the choice among energy supplies must depend in part on the extent to which matter is used up.

More generally, there are a whole host of scarcities which constrain the availability of energy supply and which fall more strongly upon some than on others. Hannon himself notes some examples of scarcities of one kind, those specific to individual supply sources. Though hydroelectricity and geothermal power rank high by Hannon's criteria, he agrees that there is simply not very much of either of these. That is, there is a scarce factor of production, hydroelectric and geothermal sites, respectively.

Specific scarcities are easy to see, and no one will be deceived. But there are more subtle scarcities. All forms of energy supply and indeed all forms of production require the uses of labor and capital, the "fund coordinates" of Georgescu-Roegen's tables. When one considers that total energy expenditures are perhaps eight percent of national income, one must conclude that payment is being made for a multitude of scarce factors among which types of energy are by no means dominant.

I also have some problems with Hannon's definitions of energy in his formulas. Note that the using up of the fuel itself is not an energy cost in Hannon's sense. Thus, in production of electricity from coal, the energy input is that used in construction of the plant and in mining and transporting the coal but does not include the coal itself. Hence the potential energy in the chemical structure of carbon is not an energy input. The using up of scarce mineral resources, which most analysts would regard as the central issue in energy economics, is not addressed at all.

For a fuller account of the "orthodox" position, let me return to Georgescu-Roegen's model, of which the production assumptions are set out in Table II and the pricing implications in Section VI. There is nothing here which is inconsistent with standard economic analysis. Let me amplify the model in ways which are thoroughly consistent with his underlying approach.

(1) The constraints on energy in situ and indeed matter in situ (ES and MS, in Georgescu-Roegen's notation) are not merely flow constraints but also fund constraints. For most forms of energy in current use, and for minerals, there is a fixed stock in existence which is used up over time. Whatever is used of fossil fuels or uranium in one period of time cannot be used later. For minerals used as materials, the same statement holds to the extent that recycling is not profitable. In a more sophisticated treatment, it must be recognized that the stocks of minerals for energy and material purposes come in many different availabilities with differing costs of capital and labor for their extraction. Hence, the scarcity of exhaustible resources is intertwined with the scarcity of capital and labor.

(2) It must also be recognized that there are alternative techniques of each kind of production, whether of energy or of goods. The alternative techniques differ in their energy-intensity (energy used per unit output). Of course, since nature supplies no free gifts, the less energy-intensive techniques will require more inputs of other kinds, capital, labor, land, or matter.

The fund constraints on certain energy sources emphasizes the importance of intertemporal choice; somehow or another, by collective action or by individual actions driven by the forces of the market, exhaustible energy and mineral resources will be allocated between the present and the future. At the same time, the decision to hold and increase the stock of capital goods also reflects intertemporal choice. In an efficient system, these choices must be consistent with each other. This has the following implication, first observed by my teacher, Harold Hotelling: the price or rent associated with the scarcity of the exhaustible resource must rise at a rate equal to the rate of interest, which measures the perceived cost of accumulating or maintaining capital. The "price" of an exhaustible resource referred is the price above and beyond the costs associated with extraction from the ground; it should ideally measure the loss to society in the future from using the scarce commodity now.

The price of exhaustible resources, such as oil, must then go up with time, relative to other prices. Point (2) above, the presence of alternative techniques of production, then implies a shift away from the use of the resources that are growing more expensive. Thus, the use of those energy and material resources which are subject to a fund constraint (exhaustible resources) will be restrained and more left available for future generations.

The growing scarcity and higher prices of exhaustible resources do indeed make us poorer (all other things being equal). That is in the nature of the case and nothing can be done about it. The scarcity is not created but only indicated by the price rise. When prices do not rise, the economy will have to adjust to the scarcity sooner or later, and it will do so in a less efficient manner.

The signalling role of prices is not confined to the special category of exhaustible resources. Scarce sites for the production of flow sources of energy and other goods, e.g., hydroelectric or geothermal locations, have scarcity rents attached to them in just the manner of agricultural land. If demand for energy rises, the prices of these sites become higher.

But it must be conceded that this nice attunement of price movements over time to growing scarcities is not realized in practice. It could only be enforced by having markets today for future energy sources, for example, a futures market in oil. Hence, the future scarcity of fossil fuels affects the present only through anticipations. The expectation of future higher prices slows current production and exploration, as indeed it should. But since anticipations are necessarily uncertain and subjective, their effect is enfeebled. If we evaluate scarcities at current prices, we may get a misleading impression of future needs. Considerations such as these underlay Georgescu-Roegen's concern over the "price complex."

Georgescu-Roegen, in common with much of the literature on energy and on long-run growth in general, concentrated on the possibility of a stationary state. By definition, a state of affairs in which exhaustible resources are being used cannot be stationary, since the fund of those resources is being diminished. Since some kinds of exhaustible resources are still in very large supply, certainly coal and probably also uranium and oil shale, the analysis of the stationary state is not immediately relevant. But one may still look ahead and see if such a state is possible. Here, Georgescu-Roegen's warnings are salutary and mesh with his concerns over the use of current prices. We may evaluate new sources of energy in ways which use current prices based on current uses of nonrenewable sources of energy. Thus, photovoltaic solar cells are produced with conventional energy sources and priced accordingly. These availabilities and prices could not persist into an era in which all energy is solar. The means of acquiring solar energy would themselves have to be produced with the aid of solar energy (including, of course, other forms of solar energy, such as biomass or wind).

The viability of a system of purely renewable energy would have to be investigated empirically, by standard feasibility methods in input-output analysis including the accompanying pricing, to which Georgescu-Roegen himself has contributed so much, along with David Hawkins and Herbert Simon. One cannot be sure of the outcome until the calculation is done, but I am a bit surprised at Georgescu-Roegen's pessimism. It seems very likely to me that even with existing techniques a world run purely on renewable energy (a stationary state) would be feasible. No doubt it would be a considerably poorer world than we have now, but we might expect technological improvements to remedy that. After all, no great breakthroughs in scientific or technological principles are needed, much less than for successful power from fusion; all that is required is a series of improvements which have quite frequently accompanied great expansions of an industry.

There is no way of avoiding Georgescu-Roegen's other objection to the stationary state, the inevitable degradation of matter. It can be balanced only by increasing economies in the use of matter. It may be that the vast reduction in material inputs to computers may hold the key; information may substitute for goods embodying a great deal of material.

References

1. See, e.g., N. Georgescu-Roegen, "Energy Analysis and Economic Valuation," Southern Economic Journal, 45 (1979): 1023-1058.

Bruce Hannon

5. Reply

Professor Arrow wonders why I made the energy cost calculations. I had hoped that I had clearly explained that I seek to determine the feasibility of an energy transformer (does it produce more of a desired energy form than could be made from its processing energy?) and the desirability of the transformer (is this net production greater or less than competing transforming technologies can yield?). I do believe that energy is an important accounting variable on which to uniquely evaluate the feasibility and desirability of energy transformers. It can be shown that if the energy discount rate is higher than the financial discount rate (and it appears to be) nothing can be said about energy feasibility knowing only that a transformer is financially feasible.

My analysis of energy scarcity is explicit, including a graph showing when transforming technologies would be abandoned. It is this procedure which allows me the point of view that the input fuels can be ignored in the energy calculation. I see the input fuels as arising from their natural state while the processing energies have already been produced as a surplus, ready for investment or consumption. These input energies are part of the stock of natural low entropy endowments which we tap as we need and have the technologies to do so. One might argue that to ignore the input fuels is to ignore the alternate uses to which these fuels might be put. An economist might argue that inputs of sunlight should be ignored but not inputs of coal. But sunlight has an opportunity cost. It is a part of the weather system and to shift its use is to deny its present use. All of the inputs have some form of alternate use but assigning value to the use is terribly complicated. It might vary from zero to a substantial value, over long periods of time. I have chosen to ignore the value of the input fuels beyond their extraction cost. One might try to examine the many ways in which these fuels could be used to produce or reduce energy in a given economy and choose the largest of these values as the value of the input fuel.

Energy in the fossil or fissionable fuel form is one of the kinds of our endowment of low entropy stocks. The ores of materials such as iron, aluminum and copper and the natural materials such as limestone and marble are other such endowments. I have not been very interested in the consumption of these forms because of the relatively small amounts of low entropy embodied in their annual use. In 1967, the consumption of all such minerals amounted to about one quad while energy release was about 67 quads. The endowed low entropy flow, solar energy, was captured at the rate of about 4-5 quads in that year. Clearly the dominant draft on the endowment, other than the solar heating of the earth, was in the form of energy consumption. It is this view which leads me to think of production labor and capital as mechanisms for embodying our natural endowments in forms more suited to our present ideas of usefulness. Labor and capital are made from these natural endowments and therefore can never be considered more scarce than the endowments themselves.

Professor Arrow says that energy costs are only 8 percent or so of the GNP, but the Electric Power Research Institute cites current energy use at 16 percent of GNP and growing very rapidly (1). These dollar measures of energy may not always be terribly useful, as we noted in 1973. Professor Arrow goes on to say that energy prices (beyond cost) must rise at a rate equal to the interest rate. This clearly did not happen in 1973-74. Energy prices rose much faster than the interest rate. The connection between a resource rent and the rate of interest stated by Professor Arrow is theoretically true only if there is no effect of the scarcity on the final output of the economy, and/or if the marginal extraction cost of the resource is zero. The rent value of a resource can be broken into two independent parts: relative or Ricardian rent and absolute or scarcity rent. Ricardian rent for energy obtains from the relative thermodynamic qualities (the relative net free energy measures) of the combination of employed energy resources. Scarcity rent for energy arises from the absolute level of this energy measure for the aggregate of these energy resources. Ricardian rent tends to disappear as the diminishing resources become more thermodynamically uniform. Scarcity rent tends to disappear as the aggregate thermodynamic quality of the energy resources approaches that of sunlight.

If we imagine for a moment that an international futures market in energy did exist in 1973-74, then this rapid price rise implies that the energy discount rate was negative. It is the denial of the existence of such a negative physical rate that must have led Hotelling to his theorem. But could it have existed? (Herman Daly, in his Postscript, seems to think so.) I think not in the long run. The natural resource units likely become harder to get as they are exhausted and people do change their preferences (and therefore reduce their demand for the resource) as the resource is depleted.

Finally, Professor Arrow notes that the feasibility of solar cells is an empirical question. I certainly agree but would rather know if the solar cells could produce the energy needed in their manufacture and maintenance. Nature seems to have done so, albeit at very low energy efficiencies (N.B., I do not expect natural systems to make their own input sunlight just as I do not expect fossil units to make their own input fuel). So far, however, natural systems set the energy efficiency standard for converting and storing solar energy.

References

1. EPRI Journal, December 1980, p. 27.

Robert Costanza

6. Embodied Energy, Energy Analysis, and Economics

The thesis that available energy limits and governs the structure of human societies is not new. In 1886 Boltzmann pointed out that life is primarily a struggle for available energy. Soddy (1) stated in 1933: "If we have available energy, we may maintain life and produce every material requisite necessary. That is why the flow of energy should be the primary concern of economics" (p. 56). The flow of energy has not been the primary concern of mainstream economists, although the importance of energy to the functioning of economic systems has by now been recognized by almost everyone. Who can deny the dramatic effects of the 1973 Arab oil embargo and the 1979 Iranian revolution? The debate now focuses on the nature and details of the energy connection and the conclusions are critically important to several aspects of national policy. This paper extends earlier input-output based analysis of energy-economy linkages by incorporating the energy costs of labor and government services and solar energy inputs.

The flow of energy is the primary concern of what has come to be known as energy analysis (2, 3, 4). An important aspect of energy analysis is determining the total (direct and indirect) energy required for the production of economic or environmental goods and services. This total quantity has been termed the embodied energy. For example, the energy embodied in an automobile includes the energy consumed directly in the manufacturing plant plus all the energy consumed indirectly to produce the other inputs to auto manufacturing, such as glass, upholstery, steel, plastic, labor, capital, etc. A problem immediately apparent from this definition involves the procedures chosen to calculate indirect energy requirements. Embodied energy values are thus contingent on methodological considerations.

Input-output (I-O) analysis is well suited to calculating indirect effects in a systematic and all-inclusive accounting framework. Hannon (5) and Herendeen and Bullard (6) have adapted this technique to calculate embodied energy. Controversy still exists concerning the relevant system boundaries for such calculations (3).

Available Energy and Embodied Energy

Available energy is the ability to do work. Work is required to change the position or arrangement of matter to one which would not occur spontaneously. Thermal energy or internal energy is the internal random kinetic energy of the particles in a system. It is related to temperature and mass and measured in British Thermal Units (BTU), Calories (Cal), Joules or equivalent units. For example, a BTU is the amount of energy necessary to raise the temperature of one pound of water one degree Farenheit. Sadi Carnot pointed out that thermal energy can do no work unless a temperature gradient exists. This is the basis of the second law of thermodynamics and has led to several formulations of available energy, a quantity more related to ability to do work than thermal energy. In a general sense any kind of gradient (and not just a temperature gradient) represents available energy or potential work. Any organized structure or non-homogenous arrangement of matter will spontaneously "fall apart" or tend toward homogeneiety according to the second law of thermodynamics. An "engine" can be inserted and some of this "falling apart" can be converted into "putting together" something else--work. Since the "engine" is also an organized structure which is itself falling apart, the concept of net energy is important (7). The engine itself must require less putting together than the amount it can produce by capturing the falling apart of the source in question. Available energy is thus a property of a system or some components of a system. It is not a commodity. Oil is not available energy - oil possesses available energy because of the large chemical gradient between it and an oxidizing environment. A block of ice in Antarctica has little available energy while the same block would create a considerable amount if moved to Ecaudor. The shipment might consume more avialble energy than the increased potential, however.

Reversible thermodynamics deals with equilibrium systems - ideal cases that have great theoretical importance, but are far removed from everyday reality. The ideal Carnot engine operates at the theoretical maximum efficiency, but to do so it must operate infinitely slowly. While this rate of operation reduces wear and tear on the engine to zero and allows us to ignore engine maintenance energy, it probably won't get us to work on time!

The formal science of irreversable thermodynamics, which deals with systems far from equilibrium, is only now in its infancy (8). Irreversible systems operate at finite rates and engine maintenance energy becomes an important component in any realistic definition of efficiency.

Embodied energy accounts for the falling apart of the engine as well as the source by incorporating the entire system in a comprehensive accounting framework. It is, therefore, a potentially useful tool in the measurement of overall energy availability and system efficiency in irreversible systems. By measuring the direct and indirect energy investments required to produce economically useful gradients (goods and services) the size of the gradients may be estimated in consistent units in the same sense that dollar estimates of capital stocks are estimated from dollar investment data.

System Boundaries

The choice of system boundaries is critical because it determines the distinction between net inputs and internal transactions. Net inputs are considered to be independent and exogenously determined while internal transactions are endogenous and interdependent. The I-O technique, in essence, distributes a net input vector through a matrix of internal interactions to balance against a net output vector.

Most recent embodied energy calculations based on national I-0 tables have employed the standard I-0 boundary definitions (6). Under these definitions the net input (or value added) vector includes labor, government, capital, and energy and other natural resources (raw materials). The corresponding financial categories are employee compensation (EC), indirect business taxes (IBT) and property type income (PTI). The sum of these net inputs in dollar units is Gross National Product (GNP). Energy (fossil fuels, nuclear fuels, and solar) is a small component of GNP in dollar units. This has led several people to conclude that energy is a minor component in economic production. This conclusion would be accurate _if_ the components of the net input vector as currently defined were mutually independent, as is usually assumed.

Most proposals to increase the "energy efficiency" of economic activity are ultimately based on the assumption of mutual independence of primary factors since increasing energy efficiency entails substituting other "primary factors" (capital, labor, government services, or other natural resources) for fuel inputs. The question is: are the components of the net input vector as currently defined really independent? Are the conventional "primary factors", capital, labor, natural resources, and government services, free of indirect energy costs? A strong case can be made for the contention that they are not (7, 9, 10). This paper presents the case for the interdependence of the currently defined "primary factors", details a method for calculating embodied energies taking account of this interdependence using I-O data, and interprets the results.

Primary Factors

From a physical perspective the earth has one principle net input--solar energy. While very small amounts of meteoric matter also enter and deep residual heat may continue to drive crustal movement, there is certainly no stream of spacecraft carrying workers, government mandates, and capital structures onto the planet. Thus, unless we are colonized by superior beings from another planet, practically everything on earth can be considered to be a direct or indirect product of past and present solar energy. The same cannot be said for the other "primary" factors. Fossil fuels and other natural resources are millions of years of embodied sunlight. Environmental services are embodied sunlight of more recent origin. Humans, under this view, are the product of millions of years of solar powered R & D (natural selection), and are maintained by an agriculture which uses both current sunlight and fossil sunlight. From this perspective, industrial capital is even more obviously created by the economic process and is not a net (or primary) input.

As Georgescu-Roegen points out: "On paper, one can write a production function any way one likes, without regard to dimensions or to other physical constraints" (11, p. 97). Doing just this has allowed some economists to ignore critical real interdependencies and to conclude, for example, "There are presently extensive possibilities of substitution between resources and other factors (capital)" (12, p. 64). Georgescu-Roegen goes on to say: "In actuality, the increase of capital implies an additional depletion of resources." Odum (10) has pointed out that the currently defined "primary factors" are really interdependent by-products of our one real, observable net input--solar energy.

How can this interdependence of primary factors be taken into account in an analytical model? In an I-O framework, one can simply expand the foundaries so that the net input to the analytical system coincides with the net input to the real system. In practice, most of the interdependencies can be captured by considering households and government to be endogenous sectors. This represents a return to Leontief's original concept of a "closed" economic system (13). The thermodynamic definition of a closed system is more appropriate in this case, however, since the system boundaries are placed so that only current solar energy and the energy embodied in fuels and other natural resources enters as a net input.

In a closed Leontief model households and government are treated like any other sector, with technical coefficients based on the household and government consumption (inputs) used to produce labor and government service outputs. As with standard input-output analysis this is strictly an accounting of inputs and outputs with no normative implications. The question of whether the current standard of living and level of government spending is good or bad, too high or too low, necessary or wasteful is not and need not be asked in this format.

Input-Output Based Energy Accounting

The input-output technique for calculating embodied energy involves defining a set of energy balance equations (one for each sector) and solving the resulting set of simultaneous linear equations for the energy intensity coefficient vector $\underline{\varepsilon}$, which is the energy required directly and indirectly to product a unit commodity flow.

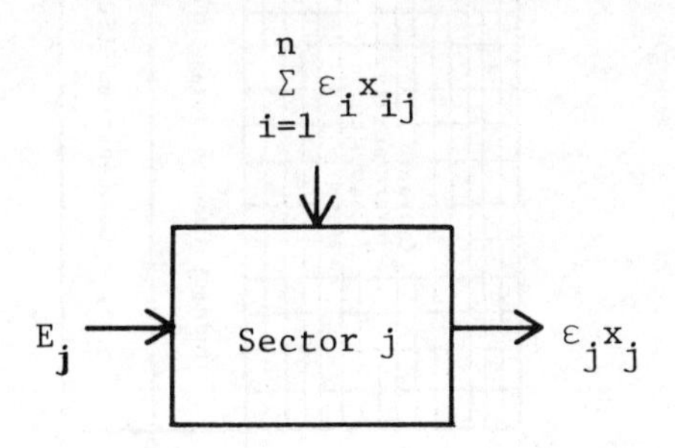

Figure 1. Energy balance for sector j.

Figure 1 shows the basic energy balance for a sector, where

x_{ij} = transaction from section i to sector j

x_j = total output of sector j

E_j = external direct energy input to sector j

ε_j = embodied energy intensity per unit of x_j

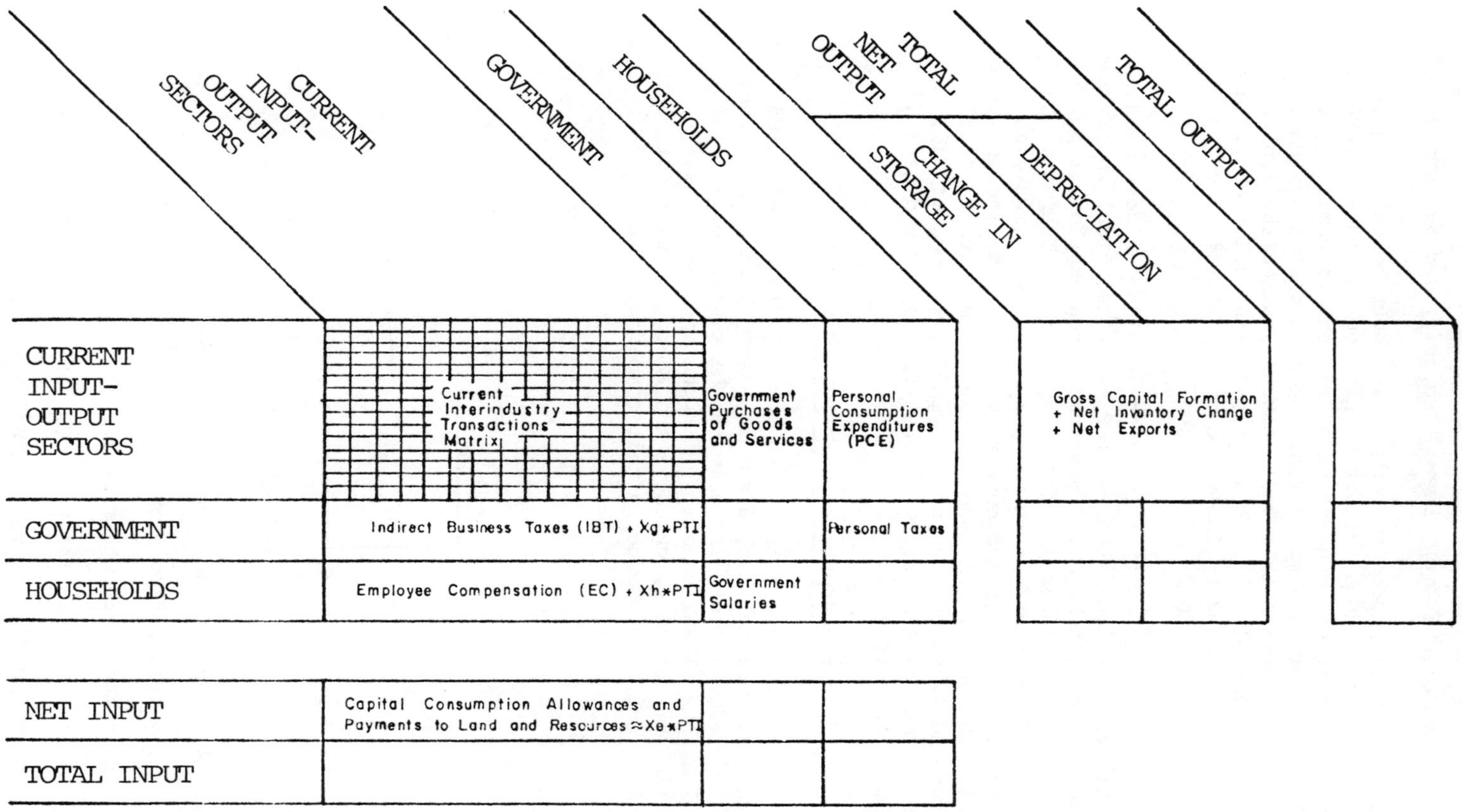

	CURRENT INPUT-OUTPUT SECTORS	GOVERNMENT	HOUSEHOLDS	TOTAL NET OUTPUT: CHANGE IN STORAGE	TOTAL NET OUTPUT: DEPRECIATION	TOTAL OUTPUT
CURRENT INPUT-OUTPUT SECTORS	Current Interindustry Transactions Matrix	Government Purchases of Goods and Services	Personal Consumption Expenditures (PCE)	Gross Capital Formation + Net Inventory Change + Net Exports		
GOVERNMENT	Indirect Business Taxes (IBT) + Xg*PTI		Personal Taxes			
HOUSEHOLDS	Employee Compensation (EC) + Xh*PTI	Government Salaries				
NET INPUT	Capital Consumption Allowances and Payments to Land and Resources ≈ Xe*PTI					
TOTAL INPUT						

Figure 2. Summary of modifications to the current national input-output conventions.

Thus the energy balance for the jth sector is

$$E_j = \varepsilon_j x_j - \sum_{i=1}^{n} \varepsilon_i x_{ij}$$

In matrix notation for all sectors

$$\underline{E} = \underline{\varepsilon}(\underline{\underline{\hat{x}}} - \underline{\underline{x}}) \quad (2)$$

where

$\underline{E}$ = a row vector of direct external energy inputs

$\underline{\underline{\hat{x}}}$ = a diagonalized vector of gross output flows

$\underline{\underline{x}}$ = the n x n transactions matrix

$\underline{\varepsilon}$ = a vector of total (direct plus indirect) energy embodied in a unit of outflow

one can solve for $\underline{\varepsilon}$ as

$$\underline{\varepsilon} = \underline{E}(\underline{\underline{\hat{x}}} - \underline{\underline{x}})^{-1}$$

More detailed expositions of the technique and examples can be found in (6), (14), and (15).

The modifications to include labor and government in the model involved expanding the transactions matrix ($\underline{\underline{x}}$) to include two more sectors, households and government. These sectors received goods and services from the other sectors in proportion to personal consumption expenditures (PCE) and government expenditures. They provide services to the other sectors in proportion to employee compensation (EC), indirect business taxes (IBT) and some percentage of property type income (PTI). Figure 2 summarizes the modifications and approximations, which are discussed in greater detail in (15) along with the method for estimating solar energy inputs based on land and water use. Note that with an endogenous household and government sector the GNP as currently defined is no longer the net input and output of the model. Personal consumption and government expenditures are now internal transactions leaving gross capital formation, net inventory change and net exports as the new net output. Likewise, wages and taxes are internal payments leaving capital consumption allowances and payments to land and resources as the new net input. To be complete, gross capital formation should include human and government capital formation as well. This implies that some consumption categories in the present model (such as spending on education) would be better handled as capital formation categories. Data on these categories for the U.S. economy have recently been calculated by Kendrick (16) but have not been incorporated in the present results. The effects of this omission will be discussed latter on.

Input-output methodology does not include capital stocks explicitly since it assumes static equillibrium. The flows produced by stocks and the flows necessary to maintain and expand stocks are included (as gross capital formation), so stocks are implicitly taken into account. However, even this picture is somewhat distorted. This is because by convention gross capital formation is credited to the industries producing the capital, not those utilizing it. This distortion is correctable and modifications of the I-O model giving a more accurate picture of capital flows have been performed (17). These modifications were not included in the results presented here. They have been shown to cause only a 7-8 percent change in the average energy intensity, however, (18).

Solar energy inputs were added to the $\underline{E}$ vector after correcting for the lower thermodynamic usefulness of direct sunlight compared to fossil fuel. Electricity represents an upgraded, more useful form than fossil fuel, requiring about 4 BTU fossil/BTU electricity directly and indirectly to produce (16). Likewise, fossil fuel represents an upgraded, more useful form than the solar energy which produced it. To properly account for this difference in quality, an I-O model of the bioshpere showing the complete production relations from sunlight to fossil fuel would be necessary. Since such a model does not yet exist, an approximation based on the conversion of sunlight to tree biomass to electricity in a wood burning power plant was used. This yielded a conversion factor of 2000 BTU solar/BTU fossil (18). The total solar input to the U.S. was estimated at 103×10^{18} BTU solar/yr (from data in (19) and (20)) or the functional equivalent of 51.5×10^{15} BTU fossil/yr. The solar energy was assumed to enter the economy through the agriculture, and forestry and fisheries sectors according to their relative land areas. This distribution is admittedly crude and will need to be improved as better data become available. For example, the solar input to industrial sectors via the hydrologic cycle in providing process water and carrying away wastes are not properly allocated using this approximation.

Results of Modifications to System Boundaries

The 90 sector energy input-output model maintained by the Energy Research Group at the University of Illinois was used to determine the effects of making the above modifications. The model is based on 1967 financial transactions in the U.S. economy. While physical flow I-O data would be preferred for embodied energy calculations, it is not available in the required form at the national level. Calculations using financial data are nevertheless very useful, because they yield information on the direct and indirect energy requirements to produce a dollar's worth of each of the commodities in the economy.

Table 1 lists the total primary energy intensities (in BTU fossil/$) calculated using each of four possible alternatives concerning labor, government, and solar energy inputs. Alternative A employs the conventional economic I-O boundaries and is essentially the same as the previous results of Herendeen and Bullard (6). Alternative B includes solar energy inputs, alternative C includes labor and government as endogenous sectors, and alternative D includes the modifications of alternatives B and C together.

Figures 3 and 4 show frequency plots for the four alternatives, which clearly indicate the reduction in variance of the energy intensities when labor and government are included. Including solar energy did not increase or decrease the variance significantly but this may be due to the rather crude method used in this study to estimate the distribution of solar energy to the economic sectors. A low variance indicates a more constant relationship from sector to sector of direct plus indirect energy consumption and dollar value of output.

The results were put in a regression format to highlight relationships and for significance testing. This was done by multiplying the calculated energy intensity for each sector (in BTU/$) by the sector's dollar output to yield the direct plus indirect energy input (in BTU). This was used as the independent variable with total dollar output as the dependent variable. Figures 5-8 show the results for each of the four alternatives. The primary energy sectors (1-7) were found to be outliers and the regression results excluding them are also presented. The regression lines are indicated on the plots along with the r^2 values, equations, and the t statistics on the parameters (in parenthesis below the parameter values). Table 2 lists the r^2 values, F statistics, and significance levels for each of the alternatives.

The results indicate a significant relationship between embodied energy and dollar output when labor and government are included as endogenous sectors (alternatives C and D). Several problems of omission still exist but the trend is clear. As more of the indirect energy costs are taken into

Table 1. Embodied energy intensities for 92 U.S. economy sectors in 1967 for four alternative treatments of labor, government, and solar energy inputs. (All Values in Btu fossil/$)

Sector (numbers in parenthesis are BEA sector equivalents)	Alternative			
	A Excluding Labor and Government Services Feedbacks and Solar Energy Inputs	B Excluding Labor and Government Services Feedbacks but Including Solar Energy Inputs	C Including Labor and Government Services but Excluding Solar Energy Inputs	D Including Labor and Government Services Feedbacks and Sol Energy Inputs
Coal Mining (7)	5,143,600	5,172,000	5,455,600	5,807,500
Crude Petroleum & Natural Gas (8)	2,920,300	2,929,200	3,188,600	3,469,050
Petroleum Refining & Related Products (31.01)	1,422,300	1,432,250	1,748,400	2,085,500
Electric Utilities (68.01)	505,500	513,900	796,220	1,099,950
Gas Utilities (68.02)	809,380	816,400	1,109,700	1,421,000
Other Agricultural Products (2)	81.567	775,090	381,090	1,385,400
Forestry & Fishery Products (3)	62,565	23,297,500	337,420	23,861,500
Livestock & Livestock Products (1)	55,276	340,710	363,800	1,053,500
Agricultural, Forestry & Fishery Services (4)	32,697	202,265	336,640	826,300
Iron & Ferroalloy Ores Mining (5)	65,904	87,840	395,620	755,500
Nonferrous Metal Ores Mining (6)	61,037	99,605	406,060	800,800
Stone & Clay Mining & Quarrying (9)	97,477	109,420	417,630	760,900
Chemicals & Fertilizer Mineral Mining (10)	59,002	71,645	352,820	667,500
New Construction (11)	54,804	230,245	389,770	913,950
Maintenance & Repair Construction (12)	42,803	102,060	384,680	801,350
Ordnance & Accessories (13)	41,768	78,285	381,130	772,600
Food & Kindred Products (14)	52,872	346,845	390,760	1,054,700
Tobacco Manufactures (15)	30,009	197,715	469,000	1,068,550
Broad & Fabrics, Yarn & Thread Mills (16)	68,156	167,815	392,370	830,400
Miscellaneous Textile Goods & Floor Coverings (17)	67,389	125,200	384,490	775,450
Apparel (18)	38,845	295,135	371,107	974,600
Miscellaneous Fabricated Textile Products (19)	50,462	117,365	376,750	784,250
Lumber & Wood Products, Except Containers (20)	54,159	2,829,200	372,490	3,478,850
Wooden Containers (21)	39,681	1,102,550	365,030	1,766,750
Household Furniture (22)	42,521	448,550	371,210	1,119,850
Other Furniture & Fixtures (23)	50,180	248,375	378,900	919,850
Paper & Allied Products Except Containers & Boxes (24)	88,279	366,800	405,650	1,013,750
Paperboard Containers & Boxes (25)	64,123	191,625	393,530	863,800
Printing & Publishing (26)	36,051	93,360	365,790	766,700
Chemicals & Selected Chemical Products (27)	218,430	262,410	528,740	893,050
Plastics & Synthetic Materials (28)	141,730	179,180	463,370	834,300
Drugs, Cleaning & Toilet Preparations (29)	58,672	94,725	372,110	733,650
Paints & Allied Products (30)	107,100	160,680	425,290	809,300
Paving Mixtures & Blocks (31.02)	361,470	377,745	1,003,400	1,676,300
Asphalt Felts & Coatings (31.03)	246,420	295,110	639,540	1,082,750
Rubber & Miscellaneous Plastics Products (32)	95,144	128,870	432,050	814,050
Leather Tanning & Industrial Leather Products (33)	133,600	146,235	369,910	627,950
Footwear & Other Leather Products (34)	51,714	105,720	367,620	751,400
Glass & Glass Products (35)	56,916	105,600	379,200	763,400
Stone & Clay Products (36)	97,629	124,345	423,790	789,300
Primary Iron & Steel Manufacturing (37)	191,670	211,570	517,450	876,150

able 1, continued

Sector 'numbers in parenthesis are BEA sector equivalents)	Alternative A Excluding Labor and Government Services Feedbacks and Solar Energy Inputs	B Excluding Labor and Government Services Feedbacks but Including Solar Energy Inputs	C Including Labor and Government Services Feedbacks but Excluding Solar Energy Inputs	D Including Labor and Government Services Feedbacks and Solar Energy Inputs
·imary Nonferrous Metals Manufacturing (38)	61,002	91,975	365,500	712,200
·tal Containers (39)	169,170	196,040	503,020	877,200
·ating, Plumbing & Fabricated Structural Metal Products (40)	75,663	101,485	405,180	774,200
·rew Machine Products, Bolts, Nuts, etc. & Metal Stampings (41)	72,656	98,910	404,660	776,950
·her Fabricated Metal Products (42)	68,788	97,470	391,620	756,450
gines & Turbines (43)	58,709	75,160	389,760	751,200
·rm Machinery (44)	63,300	84,800	394,780	761,600
·nstruction, Mining, Oil Field Machinery, Equipment (45)	63,345	81,015	392,830	753,550
·terials Handling Machinery & Equipment (46)	53,108	71,420	387,190	753,900
·talworking Machinery & Equipment (47)	47,631	62,730	378,670	739,250
·ecial Industry Machinery & Equipment (48)	50,064	80,210	382,800	760,000
·neral Industrial Machinery & Equipment (49)	54,503	77,855	385,670	754,350
·chine Shop Products (50)	55,938	69,680	385,170	742,450
·fice, Computing & Accounting Machines (51)	28,192	47,655	358,350	722,100
·rvice Industry Machines (52)	50,201	75,140	378,460	745,100
·ectric Transmission & Distribution Equipment & Electrical Industrial Apparatus (53)	46,727	69,030	375,410	740,400
·usehold Appliances (54)	54,215	81,855	381,740	750,200
·ectric Lighting & Wiring Equipment (55)	46,291	66,535	387,030	721,100
·dio, Television & Communication Equipment (56)	26,959	50,430	363,060	738,050
·ectronic Components & Accessories (57)	36,165	55,895	371,780	742,400
·scellaneous Electrical Machinery, Equipment & Supplies (58)	44,233	66,080	374,800	740,900
·tor Vehicles & Equipment (59)	54,469	74,160	404,860	785,600
·rcraft & Parts (60)	35,540	51,870	380,510	757,600
·her Transportation Equipment (61)	51,905	190,660	396,120	881,800
·ofessional, Scientific & Controlling Instruments & Supplies (62)	37,380	65,610	367,860	740,700
·tical, Ophthalmic, & Photographic Equipment & Supplies (63)	32,033	50,580	339,640	678,150
·scellaneous Manufacturing (64)	45,403	130,965	366,890	787,350
·ilroads & Related Services (65.01)	60,218	72,575	399,140	762,500
·cal, Suburban & Interurban Highway Passenger Transportation (65.02)	56,240	61,570	348,120	655,700
·tor Freight Trasportation & Warehousing (65.03)	80,561	89,095	422,450	784,600
·er Transportation (65.04)	85,647	105,430	484,970	918,650
Transportation (65.05)	122,630	143,910	452,230	814,700
·e Line Transportation (65.06)	132,980	143,285	468,340	825,700
·nsportation Services (65.07)	5,672	11,615	346,970	706,750
·munications Except Radio & Television Broadcasting (66)	13,571	19,640	359,400	716,650
·dio & TV Broadcasting (67)	20,722	40,165	354,050	718,500

Table 1, continued

Sector (numbers in parenthesis are BEA sector equivalents)	Alternative: A Excluding Labor and Government Services Feedbacks and Solar Energy Inputs	B Excluding Labor and Government Services Feedbacks but Including Solar Energy Inputs	C Including Labor and Government Services Feedbacks but Excluding Solar Energy Inputs	D Including Labor and Government Services Feedbacks and So[lar] Energy Inputs
Water & Sanatary Services (68.03)	68,101	86,080	425,490	812,3
Wholesale & Retail Trade (69)	29,302	43,265	411,490	814,3
Finance & Insurance (70)	17,472	30,195	364,840	737,3
Real Estate & Rental (71)	26,362	45,465	357,320	707,2
Hotels & Lodging Places; Personal & Repair Services, Except Automobile Repair (72)	41,839	58,875	359,550	705,9
Business Services (73)	23,146	43,995	339,610	689,2
Automobile Repair & Services (75)	41,209	51,365	359,280	698,2
Amusements (76)	22,217	55,395	376,110	771,5
Medical, Educational Services & Nonprofit Organization (77)	3[illegible],253	46,590	354,370	704,9
Federal Government Enterprises (78)	27,503	32,695	362,330	720,7
State & Local Government Enterprises (79)	61,893	82,185	441,310	855,8
Business Travel, Entertainment and Gifts (81)	69,697	282,015	401,410	962,1
Office Supplies (82)	49,223	152,390	373,710	814,7
Government	--	--	717,160	1,393,0
Households	--	--	358,350	738,0

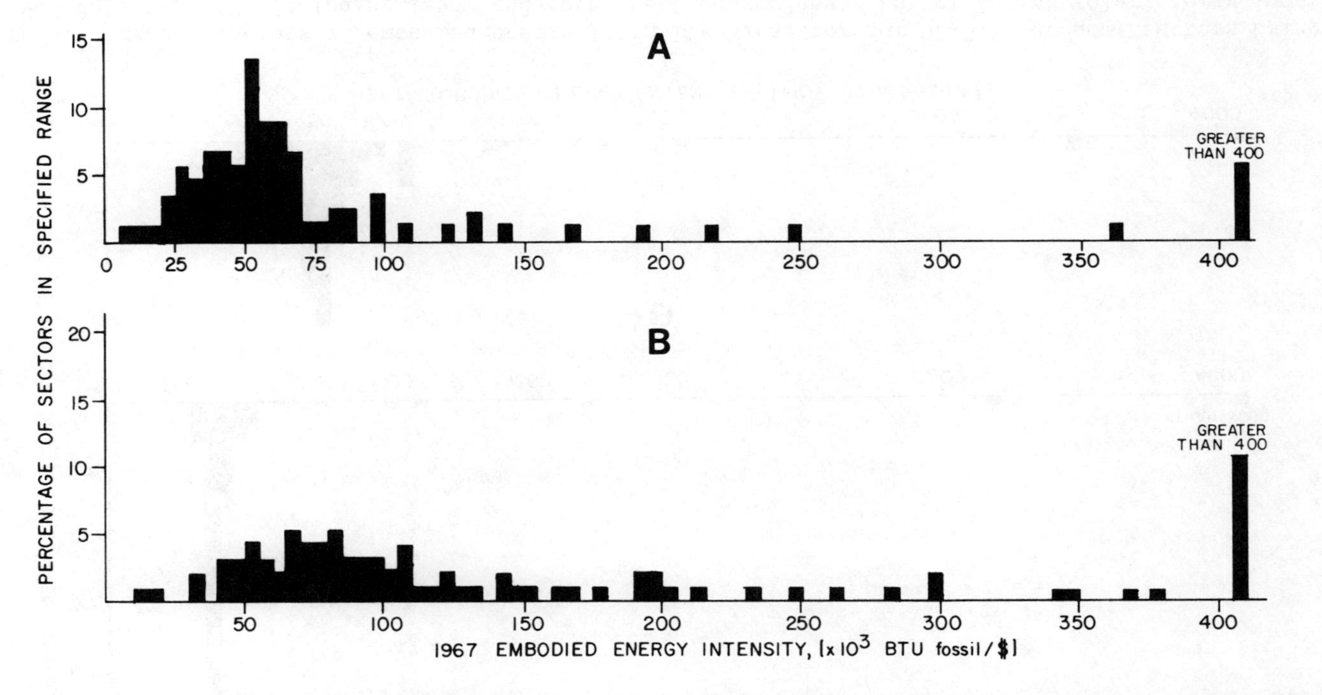

Figure 3. Frequency plots of embodied energy intensity by sector for 90 U.S. economy sectors using the 1967 data base (A) Excluding labor and government energy costs and solar energy inputs (B) Including solar energy inputs but excluding labor and government energy costs.

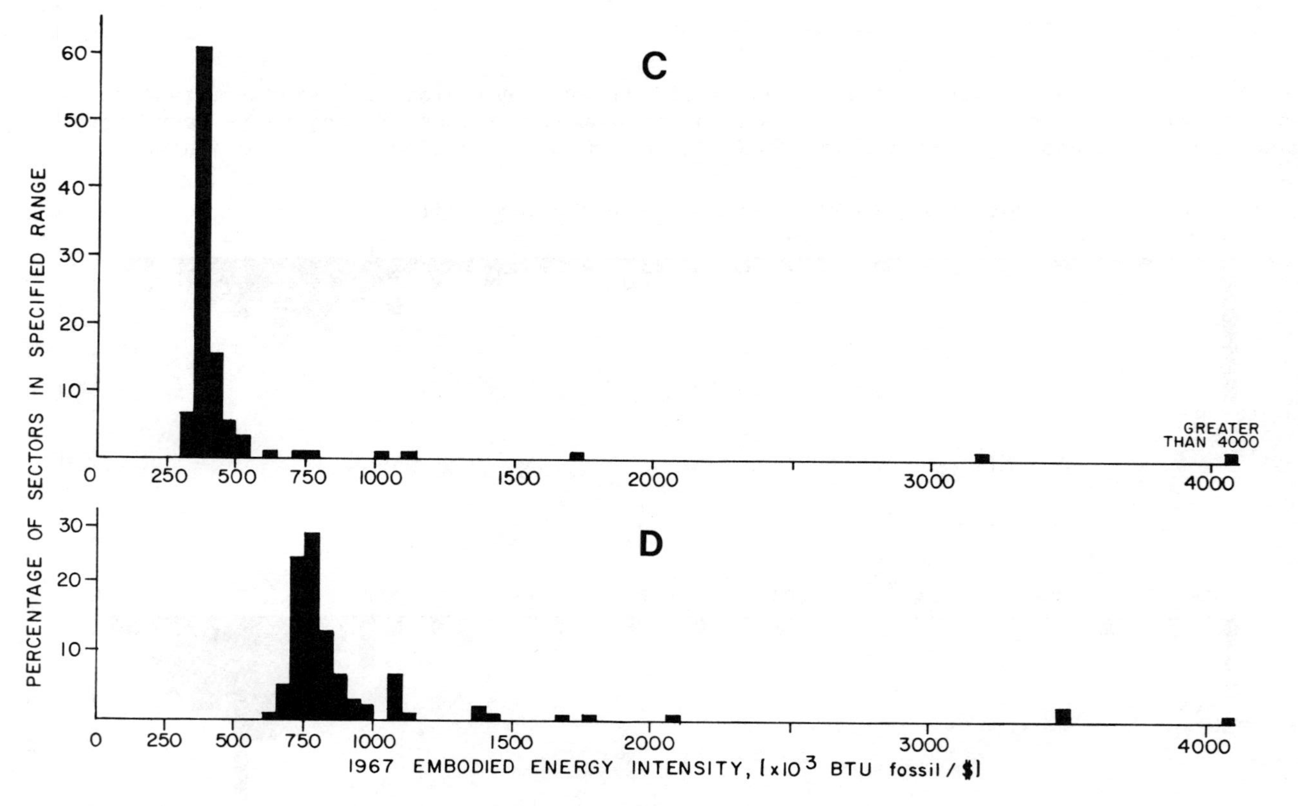

Figure 4. Frequency plots of embodied energy intensity by sector for 92 U.S. economy sectors using the 1967 data base (C) Including labor and government energy costs but excluding solar energy inputs (D) Including labor and government energy costs and solar energy inputs.

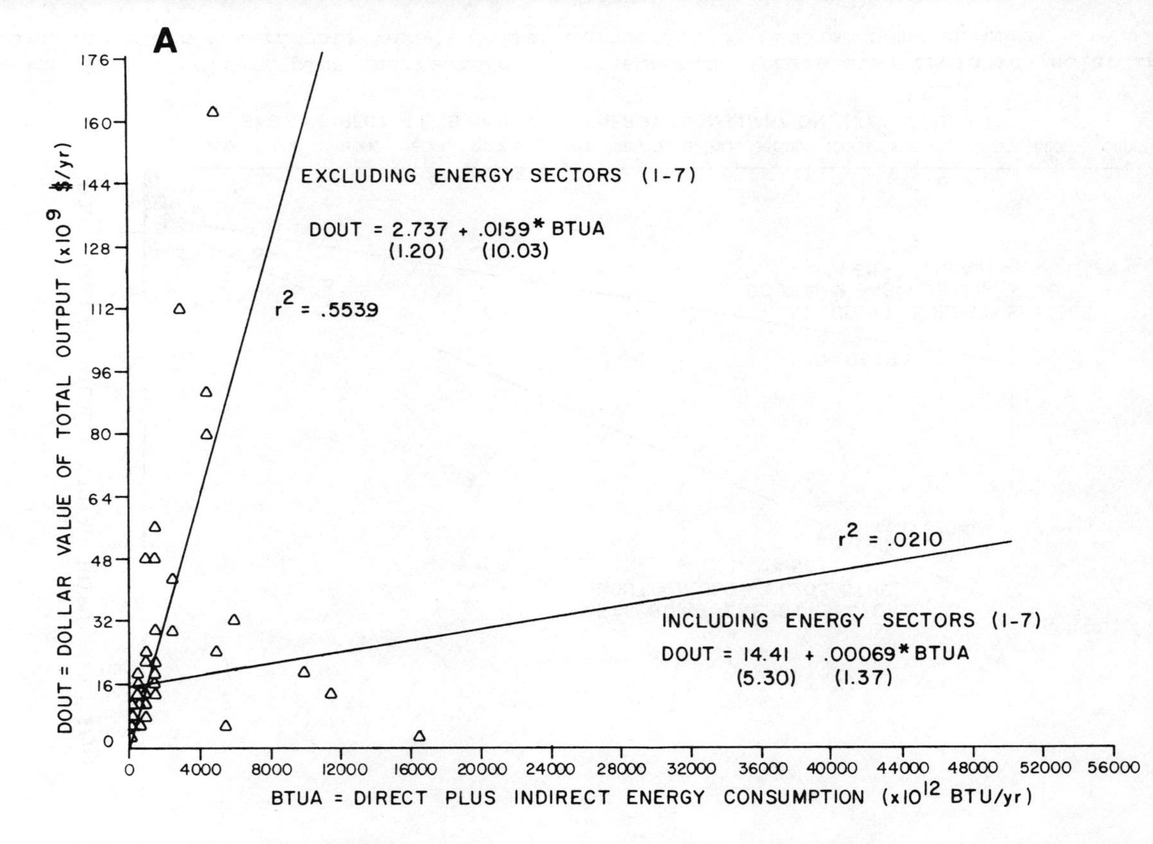

Figure 5. Plot of direct plus indirect energy consumption (calculated excluding solar inputs and labor and government) versus dollar output for 92 U.S. economy sectors.

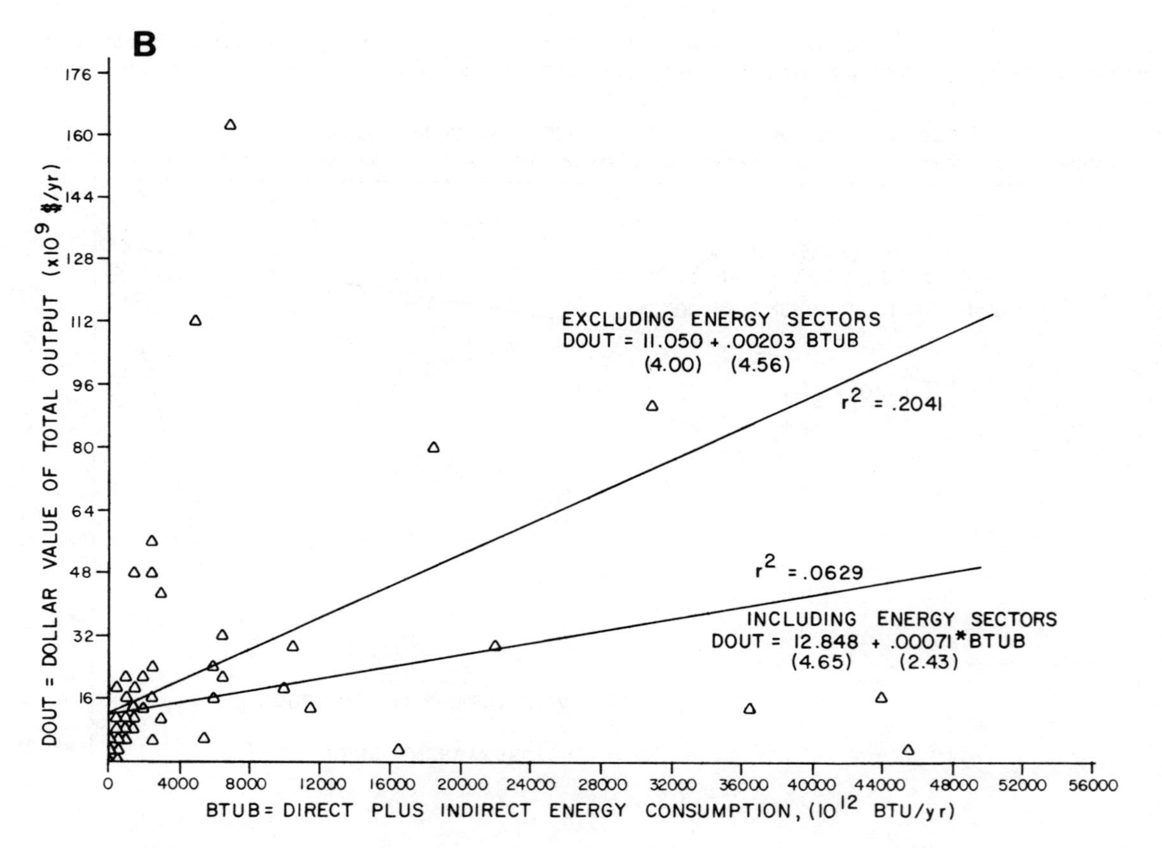

Figure 6. Plot of direct plus indirect energy consumption (calculated including solar inputs but excluding labor and government) versus dollar output for 92 U.S. economy sectors.

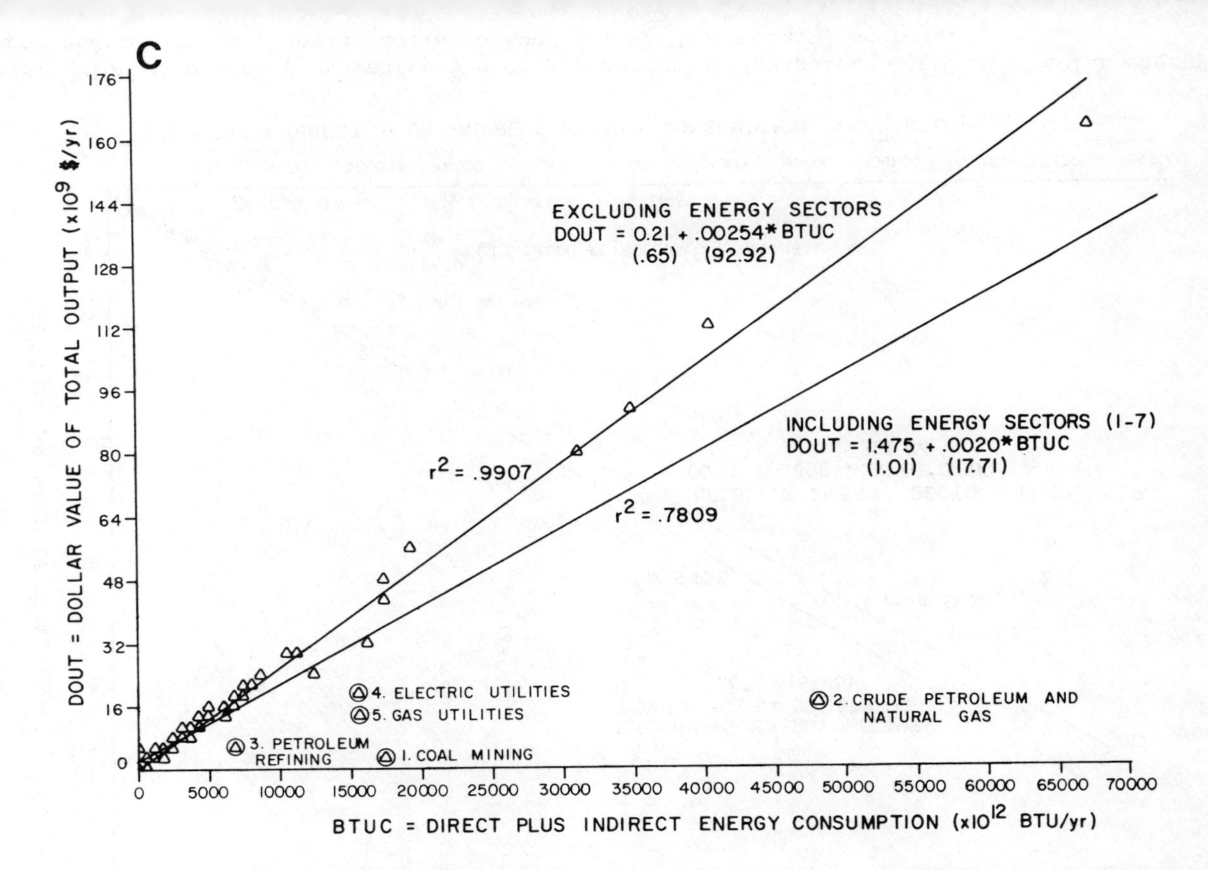

Figure 7. Plot of direct plus indirect energy consumption (calculated including labor and government but excluding solar inputs) versus dollar output for 92 U.S. economy sectors.

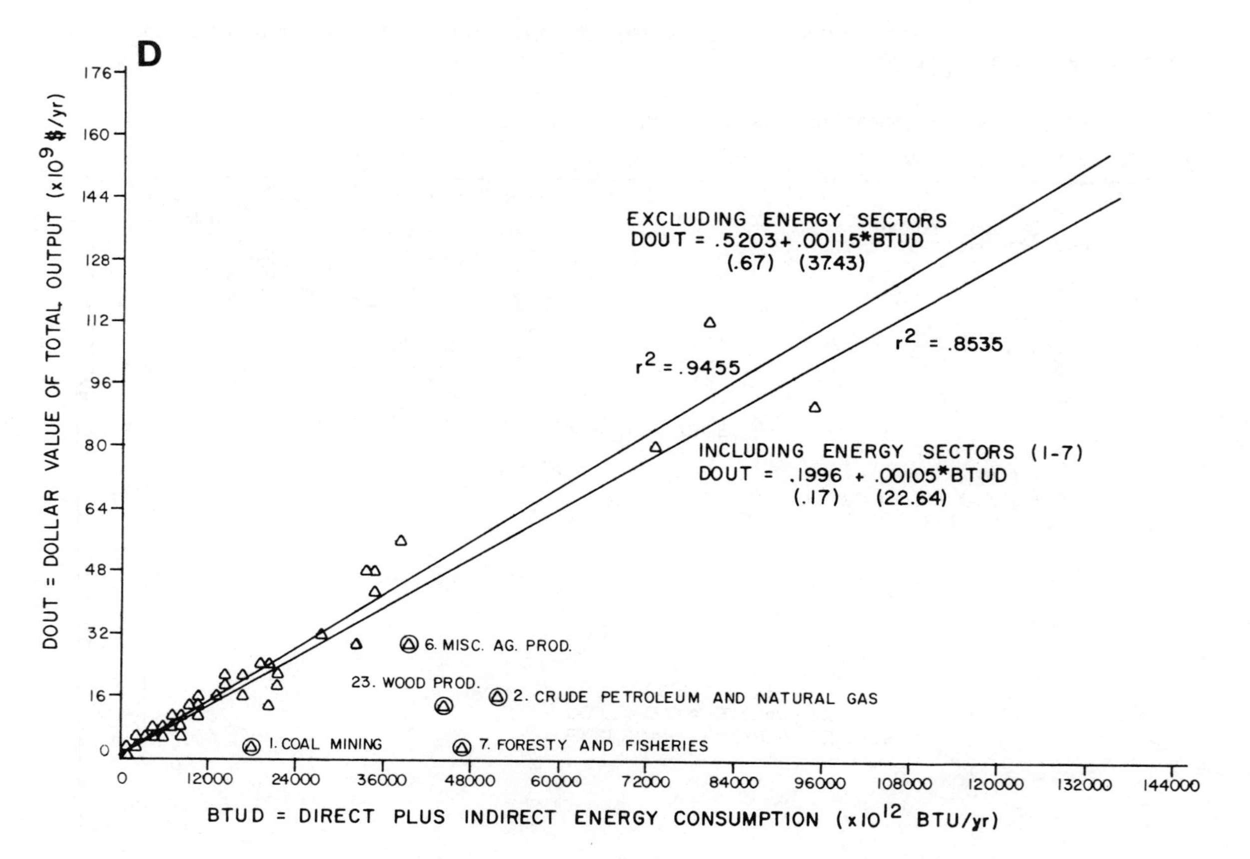

Figure 8. Plot of direct plus indirect energy consumption (calculated including solar energy inputs and labor and government) versus dollar output for 92 U.S. economy sectors.

Table 2. Regression analysis results for total (direct plus indirect) energy consumption versus total dollar output for four alternative treatments of labor, government and solar energy inputs

Alternative	Including energy sectors (Sectors 1 - 7)			Excluding energy sectors (Sectors 1 - 7)		
	r^2	F	Significance level of F test	r^2	F	Significance level of F test
A Excluding labor government and solar energy	.0210	1.89	0.1729	.5539	100.57	0.0001
B Including solar energy but excluding labor and government	.0629	5.90	0.1710	.2042	20.78	0.0001
C Including labor and government but ex-cluding solar energy	.7809	313.73	0.0001	.9907	8633.95	0.0001
D Including labor, government and solar energy	.8535	512.74	0.0001	.9454	1401.31	0.0001

account the ratio of embodied energy to dollars becomes more constant from sector to sector. The primary input sectors (sectors 1-7) are the important exceptions to this rule. Their departure from the regression line in Figures 7 and 8 is interesting and has several possible interpretations. One interpretation is that the energy intensities for these sectors are high because they represent the points of entry of available energy into the economy. Their degree of departure from the line is an indication of the net energy yield or "energy profit" they provide. In other words, their direct and indirect production costs in energy terms are much less than the energy embodied in their outputs, the difference being the amount brought into the economy from outside.

Energy/GNP Ratios

Discussions of energy and economics frequently include time series or international comparison plots of fossil, nuclear and hydro energy to GNP or GDP. The strong historical and international link between these variables is unmistakable. Several authors have suggested that "decoupling" energy and GNP is possible and would allow the economy to continue to grow while decreasing energy consumption (21, 22). The data of Herendeen and Bullard (Table 1, Column A) is often used to support this idea. If the sector to sector differences in energy intensities shown in this data were real, then it might be possible to simply shift production from high energy intensity sectors to low energy intensity sectors to lower energy use without sacrificing economic activity. To arrive at this conclusion one must accept the underlying assumption that the currently defined primary factors are independent. But because it takes available energy to produce labor and government services, capital, and other natural resources the independence assumption is not warranted. The results presented in Table 1, columns C and D, reflect the implications of an attempt to relax the independence assumption. They lead one to conclude that decoupling of energy and economic activity by simply shifting production between sectors is not a real possibility. The possibility for large changes in energy efficiency is small since, <u>all</u> things considered, total energy efficiency is fairly constant from sector to sector. Actually, the data in columns C and D lead one to conclude that energy consumption and gross capital formation plus net inventory change plus net exports are highly related, since these quantities are the net output from the economy under the revised boundary definition. GNP includes these quantities plus personal consumption and government expenditures. The extent that gross capital formation, net exports and net inventory change are seperable from GNP as a whole represents the latitude for decoupling energy and GNP. I suspect this latitude is small, especially if human and government capital formation are included as part of gross capital formation, as Kendrick (16) has suggested, and not as consumption. Inclusion of human and government capital formation would also significantly lower the mean embodied energy intensity since it would increase the redefined net output from the economy. For alternative D the mean would drop from 12.20 x 10^5 BTU fossil/1967\$ to 1.88 x 10^5 BTU fossil/1967\$ (15).

Double Counting

Slesser has commented that including labor costs in embodied energy calculations would involve double counting (22). This criticism was leveled against a specific method of including labor costs and it remains a valid criticism of that method. If one employs the conventional system boundaries to calculate embodied energy and then simply adds on the energy necessary to support labor one has indeed counted the same energy twice. In order to include the energy costs of labor without double counting, the system boundaries had to be changed. The net output with the revised boundaries does not include the support of labor which is now an internal transaction (see Fig. 2). The total energy budget is now allocated to gross capital formation, net inventory change, and net exports. The total energy requirement equals the total energy input and the energy cost of labor is accounted for. Note, however, that in any I-0 accounting system gross flows and net flows must be kept straight. One can never add up internal transactions in an I-0 table and expect them to equal net inputs or outputs. With the expanded boundaries net output and input are no longer equal to GNP but rather to GNP less labor and government transactions.

Embodied Energy Theory of Value

Several authors have proposed various forms of an energy theory of value (1, 9, 24, 25). The idea is summarily dismissed by neoclassical economists (16, 27, 28) on the grounds that energy is only one of a number of primary inputs to the production process. This dismissal is unwarranted if the traditional "primary factors" are in reality interdependent. The results presented in this paper indicate that if the interdependencies among the currently defined primary factors are real, then embodied energy values calculated taking this into account show a very good empirical relation to market determined dollar values. Herendeen (29) has shown that the necessary and sufficient condition for a constant energy intensity vector using the I-O method is that the value added vector (in dollars) and the direct energy input vector (in BTU) be proportional. If all factors other than energy are moved from the net input vector to the transactions matrix, then this proportionality is to be expected.

One might ask why we could not do the same thing we have done with energy with any of the other currently defined primary factors. We could thus support capital, labor or government service theories of value. The answer is that <u>on paper</u> we could. We must look to physical reality to determine which factors are net inputs and which are internal transactions. Can any one seriously suggest that labor creates sunlight?

An embodied energy theory of value thus makes theoretical sense and is empirically accurate once the system boundaries are defined in an appropriate way. It is really a cost of production theory with all costs carried back to the solar energy necessary directly and indirectly to produce them. The results indicate there is no inherent conflict between an embodied energy (or energy cost) theory of value and value theories based on utility. The empirical equivalence of these estimates, one from the cost or supply side and one from the benefit or demand side, supports basic economic principles grounded in optimization, while giving them a biophysical basis. Without realizing it, the flow of energy <u>is</u> the primary concern of economics.

Conclusions

The results presented here indicate that, given the appropriate perspective and boundaries, market determined dollar values and embodied energy values are proportional for all but the primary energy sectors. The required perspective is an ecological or "systems" view which considers man to be a part of and not apart from his environment. A few economists have already taken this perspective (20, 31, 32) and the implications for a new ecological economics that links the natural and soical sciences are great (23). Embodied energy may help provide such a link as an empirically accurate common denominator in ecological and economic systems. With the appropriate boundaries embodied energy values are accurate indicators of market values where markets exist. Since they are based on physical flows they may also be used to determine "market values" where markets do not exist, for example, in ecological systems. This is one way of "internalizing" all factors external to the existing market system and solving the natural resource valuation problem. From the ecological perspective markets can be viewed as an efficient energy allocation device which humans have evolved to solve the common problem facing all species--survival.

What does all this imply for national policy? Most importantly, it implies that the physical dimensions of economic activity are not separable from energy supply limitations. The universally appealing notion of unlimited economic growth with reduced energy consumption must be put firmly to rest beside the equally appealing but impossible idea of perpetual motion. To see this a comprehensive accounting which does not ignore real interdependencies must be performed. It is easy to think you can get a "free lunch" by looking only at small parts of the system in isolation. When the whole system is analyzed, however, it becomes clear that all you can do is transfer the cost of your lunch to another segment of the system.

A frequently heard argument in support of unlimited economic growth has to do with the belief that technological change will remove resource limits. As concisely stated by Chauncy Starr (34, p. 31).

> I challenge the factual validity of the assumption that resources are limited, which is at the root of the doctrine of limited expectations. I believe that the resources available for use into the distant future are not generally limited, because history testifies that advances in technology expand the availability of resources. Technology does this by providing increased efficiency in the conversion of resources to human uses, i.e., less is needed to produce more, and in the extraction of traditional resources from the biosphere, as well as by providing methods for the conversion of dormant substances into new resources.

Resources are indeed not limited - so long as we have unlimited energy to dig deeper, search more throughly, concentrate lower grade ores and, most importantly, invest in research and development. The idea of technological change as an <u>independent</u> entity, free to grow exponentially providing more and more with less and less, is Maxwell's demon in a business suit. Maxwell's demon was a proposed "pure intellegence" capable of hypothetically sorting fast molecules from slow ones and thereby defeating the second law of thermodynamics. It took several years to realize that this scheme would not work, because any real, flesh and blood demon had to be supported with, at minimum, a light source so he could see to distinguish the fast from the slow molecules. A real demon would also have to be fed, clothed, and housed, of course, and probably entertained (sorting molecules is no fun). Likewise technological change must be supported, and that support implies the depletion of additional energy and resources (anyone who believes research and development funds are unlimited should try submitting grant proposals).

While no one can deny the importance of technological change in the past, it is important to realize that it represents a real investment of energy and resources, and is not any more independent of these constraints than are other forms of capital. Maxwell's demon must be fed.

These conclusions should not be interpreted as pessimistic. Several authors, but notably Herman Daly (32), have noted the inadequacy of GNP and other yardsticks of physical economic production as measures of social welfare. Indeed, there is nothing inherently appealing about what Boulding has called the "cowboy economy", the adolescent phase of rapid, self-conscious, often painful growth.

If we are to manage our future wisely we must be aware of the physical limitations on economic activity and learn to live well within our energy budget.

Note

A brief version of this paper appeared in *Science*, vol. 210, 12 December 1980, pp. 1219-1224.

References

1. Soddy, F. 1933. Wealth, virtual wealth and debt: the solution of the economic paradox. E. P. Dutton and Co., New York. 320 pp.

2. Gilliland, M. W. 1975. Energy analysis and public policy. Science 189:1051-1056.

3. Gilliland, M. W., ed. 1978. Energy analysis: a new public policy tool. American Association for the Advancement of Science, Washington, D.C.

4. Gushee, D. E. 1976. Energy accounting as a policy analysis tool. The Library of Congress Congressional Research Service. Washington, D.C. 667 pp.

5. Hannon, B. 1973. The structure of ecosystems. J. Theoret. Biol. 41:535-546.

6. Herendeen, R. A. and C. W. Bullard. 1974. Energy costs of goods and services, 1963 and 1967. CAC Document No. 140. Center for Advanced Computation. University of Illinois, Champaign-Urbana. 43 pp.

7. Odum, H. T. 1973. Energy, ecology and economics. Ambio. 2:220-227.

8. Nicolis, G. and I. Prigogine. 1977. Self organization in non-equilibrium systems: from dissipative structures to order through fluctuations. John Wiley and Sons, New York. 491 pp.

9. Odum, H. T. 1977. Energy, value and money. In Ecosystem modeling in theory and practice. C. A. S. Hall and J. W. Day, eds. John Wiley and Sons, New York.

10. Odum, H. T. 1978. Energy analysis, energy quality and environment. In M. W. Gilliland, ed. Energy analysis: a new public policy tool. AAAS Selected Symposium 9. Westview Press, New York.

11. Georgescu-Roegen. 1979. Comments on the papers by Daly and Stiglitz. Pp. 95-105 in Scarcity and growth reconsidered. V. K. Smith, ed. Johns Hopkins Press, Baltimore. 298 pp.

12. Stiglitz, J. E. 1979. A neoclassical analysis of the economics of natural resources. Pp. 36-66 in Scarcity and growth reconsidered. V. K. Smith, ed. Johns Hopkins Press, Baltimore. 298 pp.

13. Leontief, W. W. 1941. The structure of american economy, 1919, 1929: an empirical application of equilibrium analysis. Harvard University Press, Cambridge.

14. Costanza, R. 1978. Energy costs of goods and services in 1967 including solar energy inputs and labor and government service feedbacks. Document No. 262. Center for Advanced Computation, University of Illinois, Champaign-Urbana.

15. Costanza, R. 1979. Embodied energy basis for economic-ecologic systems. Ph.D. Dissertation. University of Florida, Gainesville. 254 pp.

16. Kendrick, J. W. 1976. The formation and stocks of total capital. National Bureau of Economic Research, New York. 256 pp.

17. Kirkpatrick, K. 1974. Effect of including capital flows on energy coefficients, 1963. ERG Technical Memo 26. Energy Research Group. University of Illinois, Urbana.

18. Odum, H. T., F. C. Wang, J. Alexander, and M. Gilliland. 1978. Energy analysis of environmental values: a manual for estimating environmental and societal values according to embodied energies. Report to the Nuclear Regulatory Commission. Contract # NRC-04-77-123. Center for Wetlands. University of Florida. Gainesville, Florida. 172 pp.

19. Vonder Haar, T. H., and V. E. Suomi. 1969. Satellite observations of the earth's radiation budget. Science 163:267-268.

20. Budyko, M. I. 1978. The heat balance of the earth. Pages 85-113 in J. Gribben, ed. Climatic change. Cambridge University Press, Cambridge.

21. Stobaugh, R. and D. Yergin, ed. 1979. Energy future: report of the energy project at the Harvard Business School. Random House, New York. 353 pp.

22. Lovins, A. B. 1977. Soft energy paths: toward a durable peace. Balinger Press. Cambridge, Mass.

23. Slesser, M. 1977. Energy analysis. Letter. Science 196:259-261.

24. Cottrell, F. 1955. Energy and society: the relation between energy, social change and economic development. McGraw Hill, New York. 330 pp.

25. Hannon, B. 1973. An energy standard of value. Ann. Am. Acad. Polit. Soc. Sci. 410:139-153.

26. Huettner, D. A. 1976. Net energy analysis: an economic assessment. Science 192:101-104.

27. Langham, M. R. and W. W. McPherson. 1976. Energy analysis. Letter. Science 192:8.

28. Peskin, H. M. 1976. Letter. Science 192:9.

29. Herendeen, R. A. 1978. Unpublished proof. Energy Research Group. University of Illinois, Champaign-Urbana.

30. Georgescu-Roegen, N. 1971. The entropy law and the economic process. Harvard University Press, Cambridge. 457 pp.

31. Boulding, K. E. 1966. The economics of the coming spaceship earth. Pp. 3-14 <u>in</u> Environmental Quality in a growing economy. H. Jarrett, ed. Johns Hopkins Press, Baltimore.

32. Daly, H. E. 1977. Steady state economics. W. H. Freeman, San Francisco. 185 pp.

33. Odum, E. P. 1977. The emergence of ecology as a new integrative discipline. Science 195:1289-1293.

34. Starr, C. 1979. The growth of limits. Saturday Review 11/24: 30-31.

35. The input-output calculations on which this article is based were performed at the University of Illinois with the assistance of the Energy Research Group as part of a joint effort with the Center for Wetlands, University of Florida. The project was funded in part by the Department of Energy, Contract No. EY-76-S-05-4398 titled "Energy Analysis of Models of the United States", H. T. Odum principle investigator. I thank B. M. Hannon, M. W. Gilliland, H. E. Daly, J. W. Day, J. Bartholomew, S. E. Bayley, and R. E. Turner for reviewing drafts and providing comments. The Center for Wetland Resources, L.S.U. provided support during the preparation of this paper.

147-164

Garrett Hardin

7220
2260

7. Ending the Squanderarchy

What would the mythical Man from Mars write home if he were asked to characterize the kind of government we live under? Probably something like this: "These people say they live in a democracy, but that's just windowdressing. Nor can their system be called an oligarchy, because there are too many fingers in the pie for the *oligo:* 535 Senators and Congressmen served and advised by 18,500 aides--68 per Senator, 27 per Congressman (1). From the way everyone wastes fuel and materials I would say that King Squander is in charge. Challenged to conserve, most citizens say they are individually helpless to stop squandering. If one person were in charge we could speak of a monarchy; but *things* are in the saddle (as their philosopher Ralph Waldo Emerson said). These poor wretches sweat under a *squanderarchy*."

How did we get into this predicament? Paraphrasing Rousseau, we can say that every bad idea is born good (2). The squander imperative grew out of the idea of progress, which was given birth by the Marquis de Condorcet (1743-1794), just before he lost his life to the French Revolution. The long title of Condorcet's book speaks of the "Progres de l'Esprit humain," but the idea of spiritual (or intellectual) progress was soon displaced by material progress (3). Science and technology were the agents of progress and (in the United States at least) the existence of a free enterprise system was said to be a necessary precondition. Progress became identified with all that was open, free, limitless. The optimism of the Victorian age was well expressed by Henry George (1839-1897) in his refutation of Malthusian views of population:

> I assert that in any given state of civilization a greater number of people can collectively be better provided for

> than a smaller. I assert that the injustice of society, not the niggardliness of nature, is the cause of the want and misery which the current theory attributes to overpopulation. I assert that the new mouths which an increasing population calls into existence require no more food than the old ones, while the hands they bring with them can in the natural order of things produce more. I assert that, other things being equal, the greater the population, the greater the comfort which an equitable distribution of wealth would give to each individual. I assert that in a state of equality the natural increase of population would constantly tend to make every individual richer instead of poorer (4).

In brief, more is better. Why? For two reasons: because of the progress of knowledge in the scientific-technological-industrial revolution, and because of economies of scale. Would more be better *forever?* The way George treated this question is significant:

> Twenty men working together will, where nature is niggardly, produce more than twenty times the wealth that one man can produce where nature is most bountiful. The denser the population the more minute becomes the subdivision of labor, the greater the economies of production and distribution, and, hence, the very reverse of the Malthusian doctrine is true; and, *within the limits in which we have reason to suppose increase would still go on,* in any given state of civilization a greater number of people can produce a larger proportionate amount of wealth, and more fully supply their wants, than can a smaller number (5).

Every economy of scale ultimately peters out in a diseconomy, and every cultural revolution is finally afflicted by diminishing returns. Did George acknowledge these facts? On the basis of the passage that I have italicized above he could maintain in a court of law that he had. But it is rhetorically significant that he did *not* italicize this passage, leaving the average, not very critical reader with the impression that economies of scale will rule always. This melioristic view pervades all 565 pages of *Progress and Poverty,* rhetorically overwhelming the feeble qualification hinted at in the 15 words italicized above. In pursuit of social justice George, like most sociologists in the following century, found overpopulation unthinkable (6).

The idea of limitless, accelerating growth was welcome to the power elite of the commercial world. Growth, change, "development," spending and rapid turnover were viewed as goods without limits. Conservation, thrift and stability were bads. Paradoxically, while this non-conservative line of thought was flourishing in technology and the commercial world that benefited from it, the contrasting view of universal conservation was becoming established in pure science (on which technology and commerce ultimately depend). Curiously, the effective beginnings of conservative thought also began in France, in the work of another man of privilege who also lost his life to the Revolution. Lavoisier (1743-1794) imposed the discipline of the analytical balance on chemistry. Following his lead the world of chemistry and physics was soon ruled by the dogma that input must precisely equal output: no creation, no destruction, nothing unaccounted for. By the middle of the nineteenth century explicit statements had been made of the Law of Conservation of Matter and the Laws of Thermodynamics. All such laws--and they are not many--are generically called "conservation laws." Despite the marvellous possibilities revealed by science in its applied aspects--technology--fundamental science came to be seen as a search for "impotence principles," as the physicist E. T. Whittaker called them (7). The most profound minds of science look for the limits of the possible; they seek *closure* of the intellectual systems.

The inherent conflict between minds that worship openness and those that seek closure was largely overlooked because of social barriers between the two groups. The wheeler-dealers of the world had open minds; theoretical scientists, more cloistered and seeking no personal potency in the world of commerce and politics, had closure-seeking minds. Popular semantics favored the first group, implicitly setting up these equivalences: open=liberal=creative; while closed=conservative=timid.

We need not wonder then that the publication of Rachel Carson's *Silent Spring* in 1962 led to an avalanche of denunciations (8). The essential message of the book was this: we live in a world of real limits, to which we must adjust our demands. This theme was taken up again ten years later in the first "Report to the Club of Rome", *The Limits to Growth*. Once more those not yet weaned from the idea of progress produced a mountain of refutations, some of them quite witty (9). Yet the Establishment of Progress was beginning to quiver. By 1979 the Establishment was definitely moving: in this year a committee in the Harvard Business School brought out its report, *Energy Future*, which acknowledges that limits are real (10). The worlds of Condorcet and Lavoisier were coming together.

The union was taking place in the realm variously called "ecology" or "environment." The conservation laws of ecology are not as neat as those of physics but they are at least as important. Two in particular should be noted. First: The extinction of a species is irrevocable. (Exceptions to this are few and of no practical importance.) Second: The destruction of an ecosystem--a stable association of many species--is either totally irrevocable or, on the ordinary time scale of human history, essentially irrevocable. A tropical rainforest, once destroyed, cannot be reestablished in a few hundred years even when all the species composing such a system are still in existence. These two ecological principles counsel caution and hesitancy, in sharp contrast to the usual inference from the idea of progress which is *Innovate now, pay later* (11).

Some of the ecological changes now taking place affect the entire world. The addition of CO_2 and other vapors to the atmosphere is creating a greenhouse effect and raising the temperature of the globe. How far will this process go? We don't know. But there is at least a small probability that the process may enter into the runaway feedback mode, ultimately converting the earth into another Venus, with a surface temperature of several hundred degrees Fahrenheit. Such possibilities bring us face to face with what Edward Teller calls the Zero-Infinity Paradox (ZIP for short)--a near zero probability of a disaster of nearly infinite seriousness (12). What should be the rational man's response to a ZIP? Teller, an advocate of nuclear power plants, says *Let's take a chance*. Ecologists, almost unanimously, say *Let's not*. Unfortunately, the definitive answer to a ZIP cannot be worked out in a test-tube or with a computer.

In 1966 the ecological point of view received a powerful boost in the world outside the natural sciences with the publication of an influential essay by Kenneth Boulding, "The Economics of the Coming Spaceship Earth" (13). Boulding contrasted the wasteful "cowboy economy" promoted by the idea of progress with the conservationist ethic of the "spaceship economy" dictated by the recognition of limits. Boulding's images are vivid but not without dangers.

As interpreted by many (though not by Boulding), the spaceship image is used to justify treating the entire world as a commons (14). A necessary (though not sufficient) condition for the preservation or wise use of finite resources is that there be a sovereign power, which Thomas Hobbes (1558-1679) called a *Leviathan*. To share resources without the discipline of such power is to head toward universal ruin. Failure to make the limit of sharing coincide with the limit of sovereignty insures that the system will move into a destructive runaway feedback mode of operation (15). Given sufficient intelligence a community may be able to survive in a managed commons (16). Socialism is a managed commons which, with good planning and eternal vigilance, *may* work. But so long as a commons remains unmanaged in a world of ever increasing demands, intelligence is almost irrelevant: overconsumption and overpollution are inevitable.

A few problems are truly spaceship problems, e.g., atmospheric pollution. For such as these we may have to create a limited spaceship sovereignty, that is, a *Leviathan* constitutionally limited to the oversight of only certain resources. We should not allow mere fashion to lead us to define as "global" problems that are merely ubiquitous, for the sufficient reason that we have not yet been able to invent a global *Leviathan*, not even a limited one. The pollution of rivers and lakes is ubiquitous, but we are making progress in cleaning up Lake Erie not by defining it as part of a global commons that includes Lake Baikal and the Rhine River but by treating it as a purely local problem (where "local" includes an entire drainage system). Most people recognize that eutrophication is a local problem and so we have been spared the nonsense of "global eutrophication."

Starvation also is a local problem, one created by allowing the resident population to grow beyond the carrying capacity of the sovereign nation in which it lives (17). Unfortunately, the essentially local nature of this problem is denied by people who speak of "global hunger." In the absence of a global sovereign the phrase "global hunger" cannot give rise to practical policy recommendations.

In recent years the public has often been misled by the thoughtless repetition of the fashionable word "interdependent." A population that has grown beyond the carrying capacity of its home territory often makes demands on other populations because (it is said) "we are interdependent." At the present time populations in desperate need of food are typically increasing at two to three per cent per year, so the "solution" of sharing tends to perpetuate unidirectional *dependence*. (Some people apparently do not know what *inter-* means.) The concepts of "global" and "interdependence" are counterproductive when they foster parasitism. Parasitism is not a wise use of resources; if nothing else, it involves wasteful transportation of materials.

Returning to the local scene we note a multitude of ways in which the squandering of resources, particularly energy, is encouraged by institutional arrangements created and defended by individuals or groups pursuing their own interests. Railroads charge more to transport steel scrap destined for recycling than they do to move new iron ore. Soft drink bottlers oppose laws that would compel the recycling of bottles. The taxicab that takes you to a distant airport is, in many cities, legally forbidden to pick up a return passenger. Intercity trucks are often compelled to travel long distances empty. Industries that produce excess heat are forbidden to use it to generate some of their own electricity ("co-generation" of product and electricity). Railroad unions indulge in "featherbedding" and bureaucrats pad their rosters to create demands for larger budgets which will enhance their political power. Wasteful throwaway packaging enriches the makers of packages. For every four trees cut down for newsprint only one ends up in the news-columns, the other three being devoted to advertisements most of which most people most of the time do not look at. And so on. It is the seeming compulsiveness of all this waste that leads the Man from Mars to conclude that progress has spawned a squanderarchy.

Characteristics of a Squanderarchy

The pathological processes that led to the present state of affairs are not hard to understand. The "invisible hand" that Adam Smith uncovered in 1776 was, for a while, thought to be universal (18). Then closer observation showed that quite often politico-economic evolution is governed by a contrary sort of process: in serving his own interests *directly* the individual may go against the interests of society as a whole (and of himself indirectly, though only marginally). In such cases society is, as Herman Daly has said, kicked by an "invisible foot." We might rewrite the relevant passage in *The Wealth of Nations* thus: "Every individual, laboring to increase his revenue as much as possible, and never trying to promote the public interest when doing so would diminish his direct gain, is moved, as it were, by an individual foot that makes matters worse for society, though this is no part of his intention." The invisible foot is, of course, involved in the tragedy of the commons.

Changes in public attitudes since Earth Day, 22 April 1970, justify a certain amount of optimism about the prospects of our changing from a squanderarchy to a conserver society (19). How far we will go in this direction, and how rapidly, are questions on which there are justifiable differences of opinion. Before discussing the problems of making the transition let us look at some of the more obvious differences between the two kinds of economy. Table 1 does not pretend to be exhaustive; it should, however, be useful in provoking discussion.

A squanderarchy (line 1 of Table 1) presumes a limitless world in which the insatiable demands of consumers promote a rapid turnover of materials and a swift degradation of useful energy ("negative entropy"). By contrast, a conserver society presumes a limited world in which people try to conserve materials and energy. Squanderers, as Thorstein Veblen pointed out long ago, make a fashion of "conspicuous consumption" (20). Conservers, by contrast, quote with approval lines from the Talmud: "Who is rich? He who rejoices in his lot" (21). Conservers recommend that we learn to be content with what we have. At its extreme, the conserver ethic tries to create a fashion of "conspicuous penury." For example, some rich young women pay high prices for pre-faded, pre-patched blue jeans (line 4 of the table).

Consumption is encouraged in the squanderarchy by the vigorous promotion of credit cards and "buying on time." Conservers should discourage most forms of consumer credit, perhaps outlawing some. Vacation travel, extravagant of energy, must be strongly discouraged in a conserving society. For example, we should discontinue the recent practice of shifting a national holiday to Monday because this arrangement creates a long weekend which allows--as was intended--workers to leave home. A return to midweek holidays, by making extensive automobile journeys impossible, would reduce the consumption of petroleum energy (line 6).

Worker mobility in general will be more discouraged in a consumer society. Business managers have long favored worker mobility, finding the task of planning easier if workers can be moved around the country like chessmen on a chessboard (line 7). The economies claimed for this practice are partly, perhaps largely, spurious: most of the social costs impinge first on the worker and his family when they are moved from the old environment in which their social positions are secure into a new one in which parents and children are to some extent aliens. Part of the cost, in the

Table 1. Contrasts Between a Squanderarchy and a Conserver Society

Characteristic	Squanderarchy	Conserver Society
1. Universe assumed	Limitless	Limited
2. Basic good	Consumption	Conservation
3. Esthetic ideal	Rapid turnover a joy	Lingering enjoyment
4. Fashionable ideal	Conspicuous consumption	Conspicuous penury
5. Consumer credit	Wildly promoted	Discouraged, restricted
6. Vacation travel	Good	Bad
7. Worker mobility	A public good	A public danger
8. Labor saving devices	Virtuous	Sinful, given unemployment
9. Energy saving	A sin	A virtue
10. Damage:burden of proof	"Innocent until proven guilty"	"Guilty until proven innocent"
11. Commercial innovation	Fast	Slow
12. Objective freedom	Greater	Less
13. Psychological freedom	Less (?)	Potentially greater
14. The Seven Deadly Sins	Only sloth a sin; pride, lust covetousness, envy, gluttony and anger are virtues	All are sins (except pride and lust, in moderation?)
15. Advertising		
a. Informational	A public good	A public good
b. Seductive ("Spend!")	A public good	A public bad
c. Competitive	A public good	A public bad
16. Semantics used when Demand exceeds Supply	"Shortage"--of Supply	"Longage"--of Demand

form of increased juvenile delinquency and crime, is ultimately passed on to society at large. It is the commonization of such social costs that enables decision makers in business firms to be unaware of the true costs of the social mobility they praise and encourage.

In getting the world's work done, energy and labor are trade-offs (lines 8 and 9). In the past we have been told that every labor-saving invention is *ipso facto* good, being assured that in the long run it will increase employment. The persistence of unemployment (the full perception of which is obscured by many forms of "hidden unemployment") now makes us doubt this simple faith. The rapidly increasing cost of fossil energy is making more and more people seriously consider the desirability of reversing history, in part, by substituting muscle power for fossil fuel power. Recently the Secretary of Agriculture announced that the government will not continue to finance research aimed at developing new machines to displace human labor in agriculture. In line with the change in policy, an organization of attorneys called California Rural Legal Assistance has sued the University of California to cease channeling agricultural funds into research designed to replace men with machines (22). Almost two centuries ago English workmen, under the banner of a mythical King Ludd, tried to stop the march of the machine, only to be massacred, jailed and transported. History, which is written by the victors, said the Luddites were wrong then; will it repeat this verdict as we move into a conserver society?

Among the most striking differences between the squanderarchy of the past and the conserver society we are becoming is the attitude toward innovations in products or processes. In the past each new product put on the market has been assumed to be innocent of danger (line 10). "Innocent until proven guilty" is an old principle of Anglo-Saxon law. When a new product proved faulty or dangerous, the expense of litigation discouraged law suits. Innovation flourished. Then came the disaster of the thalidomide babies, resulting in the passage of the Kefauver-Harris amendments to the Food, Drug and Cosmetic Act (23). Thus in the year of *Silent Spring* was the assumption of Anglo-Saxon law fundamentally revised (as far as medicine was concerned), resulting in a new presumption: "Guilty until proven innocent" (24). The National Environmental Policy Act (NEPA) carried this revolution over into the environmental area beginning 1 January 1970. With ever greater frequency, proposed alterations in the environment (building of dams, filling in of sloughs, and construction of thermonuclear plants) are held up until the results of an "environmental impact study" (EIS) are examined and approved by some authority. This change in the presumption of the law unquestionably slows up commercial and technological innovation (line 11). The change is both good and bad.

What About Freedom?

Objectively, the people living under a squanderarchy are certainly freer (line 12). Subjective truth is another matter. "Freedom is the recognition of necessity," said Hegel. Perceiving the necessity of the restrictions imposed by conservation may actually make conservers psychologically more free than squanderers (line 13). But the adjustment will take time. In the interim the public outcry against the loss of freedom may slow, or even stop, the shift to a conserving society.

One of the most profound differences between conserving and squandering societies is essentially religious. In the sixth century St. Gregory the Great named the Seven Deadly Sins: pride, covetousness, lust, envy, gluttony, anger and sloth. For over a thousand years Christians viewed these sins seriously. Then came the idea of progress and its subsequent melding with commitments to innovation and consumption. As Lewis Mumford has pointed out, the "insatiability of demands" took on the status of a sacred principle, and--except for sloth--all the deadly sins were, in effect, converted into virtues because they promoted consumption (25). Ministers of religion ceased to dwell on the Seven Deadly Sins, no doubt because they subconsciously sensed the latent antipathy in their commercially oriented congregations. As we move toward a conserver society it will be interesting to see if the Seven Deadly Sins once more become fashionable topics for sermons.

As things stand now I think the Man from Mars would identify *Playboy* magazine as one of the sacred documents of our squanderarchy, not because of its photographs but by virtue of the advertisements with their incessant glorification of all forms of consumption. *Playboy* can stand as the exemplar of all advertising.

As a medium of information--a way of telling people what is for sale, where, and for how much--advertising serves a useful purpose in every sort of economy (line 15). But modern advertising has grown far beyond this simple function. The major purpose of advertising now is to seduce people into buying more than they need, and to do this it subtly glorifies every sin except sloth. When needs are satisfied, customers are badgered into changing brands. This is usually a socially trivial act. (What difference does it make whether you use *Tweedledum* toothpaste or *Tweedledee* toothpaste?) The purely competitive aspect of advertising is largely a social bad, since it increases waste.

In 1978, the cost of advertising in the U.S. was $44 billion. Because this sum was only two percent of the Gross National Product (26) we might be tempted to dismiss it as trivial. But before doing so we should ask, How much wasteful consumption was fostered by this two percent of the GNP? There is no objective answer to this question because "wasteful" implies standards, on which opinion is divided. Nevertheless, it is probable that most people who are committed to the conserver ethic would estimate that the waste caused by advertising far exceeds two percent of the GNP.

Of much greater importance is the bearing of advertising on human freedom. Television is the most important medium of information in the lives of Americans, "information" being used in the widest sense. There are 168 hours in the week. Let us assume that 40 of these hours are used for work, 56 for sleep, 8 for necessary local transportation (to offices, schools and shops), and 10 for eating. Subtracting these 114 hours from the weekly total leaves a supply of "disposable time" of 54 hours per week. According to A. C. Neilsen, the average American in 1978 spent 30.4 hours per week viewing television (27). That figures out at 56 percent of the disposable time. Most of the programs watched are commercial programs. Considering the great skill with which TV advertisements are constructed and the pervasiveness of their messages, how free are people, really, to do their own thing (28)? We still thrill to John Milton's praise of free speech and condemnation of censorship: "Let Truth and Falsehood grapple; who ever knew Truth put to the worse in a free and open encounter?" (29) But is it a "free and open" encounter when an unknown but certainly quite small sum spent on denouncing the Seven Deadly Sins is opposed by $44 billion spent in praising six of them?

We have gone far beyond Milton. Many psychologists, notably B. F. Skinner, have convinced us that--to a large extent--we are what we are conditioned to be. The key issue in freedom is this: Who controls the conditioning process--and who controls the controller? This is a deep and subtle problem (30). It is central to the problem of changing from one type of society to another--or not changing.

Is Change Possible?

The preceding analysis has implicitly assumed that a historical change of the following sort is possible:

Squanderarchy ⟶ Conserver Society (1)

But is it? We can find many reasons for doubting this. For one thing a large and influential literature has treated perpetual growth as the only conceivable--or only moral--state of affairs. A civilization that has scoffed at the fatalism of traditional societies has created a fatalism of its own, a "Grow or Die" fatalism. In the depths of the Great Depression, Stuart Chase wrote:

> Abundance demands no compromise. It will not operate at half speed. It will not allow retreat to an earlier level and stabilization there. Pharaoh did not tell the Nile what to do; the Nile told Pharaoh what to do. The industrial discipline must be accepted--all of it--or it must be renounced. The only retreat is back one hundred years to the Economy of Scarcity (31).

The unacceptability of restraint was underlined by Nicholas Rescher a generation later when he said: "An *economy of scarcity* is, by definition, one in which justice. . .cannot be done, because there is not 'enough to go around'. . ." (32).

Despite the conservation literature of the past two decades fatalism is still evident in the writings of influential commentators of today. Consider, for instance, Irving Kristol:

> There is far too much easy and glib talk these days about the need for Americans to tighten their belts, accept a reduction in their living standards, even resign themselves to an economic philosophy of no-growth. It is dangerous and irresponsible talk. Yes, of course, the American democracy can cope with a *temporary* cessation of economic growth, as it has done in wartime. But only if it is perceived to be temporary. What few seem to realize is that a prospect of economic growth is a crucial precondition for the survival of

> any modern democrary, the American included. . .It is only so long as economic growth remains a credible reality that democracy will remain an actuality (33).

The interest vested in promoting perpetual growth is enormous. The most important fraction of this interest is probably that of the $44 billion advertising industry. It is, therefore, of considerable interest to note that many influential members of this segment are seriously examining the role they might play in a world devoted to conservation rather than to squandering. At a conference on "The Conserver Society" held at the University of Texas in November 1979, and sponsored by the American Marketing Association, many leaders of the industry expressed confidence in the adaptability of their art to the needs of conservation. The skills of "marketing" (which includes advertising) can, they said, promote conservation as well as consumption. But one thing bothered them: who would pay marketers to persuade people to *not* consume? (34).

It is not only marketers who must be paid to facilitate conservation; so also must many other people, and it is not obvious how this is to be done. The long term trend toward bureaucracy may seem to imply that a no-growth (or low-growth) conserver society will be a coercive society run by bureaucrats, but Charles L. Schultz suggests another possibility: a society that makes better use of market mechanisms to serve the public interest. Market-like arrangements can "minimize the need of coercion as a means of organizing society. . .A market approach. . .stresses incentives, not rights and duties. People or firms act in certain ways because their self-interest dictates doing so, given the existing set of incentives." (35)

As the comic-strip character *Pogo* says, we are faced with overwhelming opportunities. If the inventive genius so evident in science and technology during the two centuries of "Progress" can be transferred to the realm of politics and economics we may be able to make a moderately smooth transition along the line indicated by Eqn.(1) above. But if our abilities prove unequal to this task, the transition may take a different course:

$$\text{Squanderarchy} \longrightarrow \text{Poverty \& Chaos} \longrightarrow \text{Conserver Society} \quad (2)$$

We may have to pass through a new Dark Ages before emerging into the light of a sensible, comfortable, conserving society. Between the beginning and the end there may have to be a painful political discontinuity.

If I were betting on the future where would I place my bet--on passage (1) or passage (2)? If I were the Man from Mars I daresay I would bet on the second passage. But, as a participant observer of the human comedy, I confess that I will be satisfied with myself only if I do what I can to promote the first passage.

Population and Prosperity as Trade-Offs

It is part of the conventional wisdom of our time to suppose that a no-growth economy is inherently unstable--this despite the fact that most human societies, most of the time, have lived under no-growth conditions. As Kristol worded the unexamined assumption of our day: "It is the expectation of tomorrow's bigger pie, from which everyone will receive a larger slice, that prevents people from fighting to the bitter end over the division of today's pie."

No doubt there is much truth in this, yet we cannot forget the Talmud: *Who is rich? He who rejoices in his lot.* We need to be quite explicit about the possible roads to wealth.

The definition of wealth can be given as an equation:

$$\text{Wealth} = \frac{\text{Supply}}{\text{Demand}} \qquad (3)$$

The technique of dimensional analysis, as developed by physicists, can be applied to this. "Supply" has the dimension of material things, the sort of things listed in handbooks of resources (multiplied, of course,by a fractional factor standing for the technological "state of the art"). Demand, by contrast, in both its origins and its enduring meaning, is psychological in its dimension. Wealth is therefore also psychological in dimension--as the Talmudist (and many others) realized centuries ago. *No technological advance can take wealth out of the psychological dimension.*

There are two ways for an individual to become rich: by increasing the supply (if he can), or by decreasing his demand. If the supply cannot be increased the wise person seeks to decrease his demands (expectations, aspirations, desires, wants--call them what you will). Such is reality for the individual.

When we consider a population of individuals the wealth equation needs to be expanded:

$$\text{Wealth} = \frac{\text{Supply}}{(\text{Per capita demand}) \times (\text{Size of population})} \qquad (4)$$

With a limited supply, wealth can be increased by either of two routes: lowering the expectations of the average individual, or decreasing the number of individuals. The two forms of demand control can be combined.

In a world of limits we can become wealthy only if we subject ourselves to the discipline of demand control. This will not be easy for many reasons, not the least of which is a semantic one. Confronted with a painful discrepancy between supply and demand, the prisoners of a squanderarchy invariably speak of a *shortage* of supply. Why do they never speak of a *longage* of demand? Logically, one expression is as apt as the other: why do we always choose the first?

It must be admitted that only the word "shortage" can be found in the dictionary. The earliest usage recorded in the Oxford English Dictionary dates from 1868. The word "longage" seems not to have been coined until 1977 (36). It is easy to see how natural selection favors egotistical organisms that make demands on their environment, organisms which (if they can speak) readily complain of "shortages." It is hard for selection to favor organisms that are willing to curb their demands, in competition with organisms that will not. Demand control, though recommended by gurus for millenia, has not been the practice of the masses. Competition has not favored demand control. Yet this is precisely what a conserver society must achieve if it is to be happy in a limited world.

Confronted with an unconventional way of looking at common realities the human mind has a wonderful ability to quietly and instantly repress language that threatens egotistical impulses. Boulding's attempts (in 1945 and 1949) to get economists to reexamine their world-view of "throughput" and "stock" in terms of consumption and conservation, produced (he reports) no response whatever for more than twenty years. Not until his classic paper of 1966 did his ideas get the attention they deserved. The shift was no doubt attributable in part to a change in the intellectual climate; but it may also be that a novel idea takes several decades to mature to the stage in which it is visible to the masses.

To speak of "shortage" is to predispose the mind to look only for ways to increase supply. By speaking of "longage" we force our minds to consider the possibility of decreasing demand. "Shortage" is the semantics of a squanderarchy, "longage" the language of a conserver society.

If the world is indeed limited, conservation is required, but the level of living at which conservation is practiced is theoretically a matter of choice. A community can opt for the largest possible population; in this case individuals will be happy only if they succeed in minimizing their per capita demands. Alternatively, a community can opt for population control; if population is kept at a low level, the per capita demands (desires, aspirations, standards) can be correspondingly high. Population control offers the *possibility* of achieving psychological wealth without a policy of penury. Of course we do not yet know how to control population by "acceptable" means because the word "acceptable" has psychological dimensions that are as yet unexplored. Therein lie the problems of population and prosperity.

References and Notes

1. D. Farney, *Wall Street Journal,* 18 Dec. 1979, 1.
2. The paraphrase is of the opening sentence of J. J. Rousseau, *The Social Contract* (1762): "Man is born free, and he is everywhere in chains."
3. J. B. Bury, *The Idea of Progress* (Macmillan, New York, 1932); F. J. Teggart, Ed. *The Idea of Progress* (Univ. California Press, Berkeley, 1949); A.-N. de Condorcet, *Sketch for a Historical Picture of the Progress of the Human Mind* (Noonday Press, New York, 1955).
4. H. George, *Progress and Poverty* (Schalkenbach Foundation, New York, 1962), p. 141.
5. *Ibid,* p. 149.
6. G. Hardin, *Science* 171, 527 (1971).
7. E. T. Whittaker, *Proc. Roy. Soc. Edinburgh* 61, 160(1942)
8. R. Carson, *Silent Spring* (Houghton Mifflin, Boston, 1962); F. Graham, Jr., *Since Silent Spring* (Houghton Mifflin, Boston, 1970).
9. D. H. Meadows, D. L. Meadows, J. Randers and W. H. Behrens, III, *The Limits to Growth* (Universe Books, New York, 1972); H. S. D. Cole, C. Freeman, M. Jahoda and K. L. R. Pavitt, Eds., *Models of Doom* (Universe Books, New York, 1973); N. Macrae, *Saturday Review/World,* 20 April 1974, p. 6; M. J. Roberts, *The Public Interest,* No. 31 (1973).
10. R. Stobaugh and D. Yergin, Eds., *Energy Future* (Random House, New York, 1979).
11. F. J. Dyson, *Bull. Atomic Scientists* 31(6), 23 (1975); P. R. Ehrlich, *Bull. Atomic Scientists* 31(7), 49 (1975).
12. G. Hardin and C. Bajema, *Biology: Its Principles and Implications,* 3rd ed. (W.H. Freeman, San Francisco, 1978), p. 702.

13. K. E. Boulding in *Environmental Quality in a Growing Economy,* H. Jarrett, Ed. (Johns Hopkins Press, Baltimore, 1966), p. 3.
14. G. Hardin, *Science* 162, 1234 (1968).
15. G. Hardin, *BioScience* 24, 561 (1974).
16. G. Hardin and J. Baden, Eds., *Managing the Commons (W.H. Freeman, San Francisco, 1977).*
17. S. L. Flader, *Thinking Like a Mountain* (Univ.of Missouri Press, Columbia, 1974); G. Hardin,*The Limits of Altruism* (Indiana University Press, Bloomington, 1977).
18. A. Smith, *The Wealth of Nations* (modern Library, New York, 1937), p. 423.
19. E. F. Schumacher, *Small Is Beautiful* (Harper Torchbooks, New York, 1973); H. N. Woodward, *Capitalism Can Survive in a No-Growth Economy* (Brookdale Press, Stamford,Conn., 1976); D. L. Meadows, Ed., *Alternatives to Growth--I: A Search for Sustainable Futures* (Ballinger, Cambridge, Mass., 1977); W. Johnson, *Muddling Toward Frugality* (Sierra Club, San Francisco, 1978); K. Valaskakis, P.S. Sindell, J. G. Smith and I. Fitzpatrick-Martin, *The Conserver Society* (Harper & Row, New York, 1979).
20. T. Veblen, *The Theory of the Leisure Class* (Macmillan, New York, 1899).
21. The Pirke Aboth, in W. Durant, *The Age of Faith* (Simon & Schuster, New York, 1950), p. 363.
22. *Los Angeles Times,* 1 February 1980, Part I, p. 3; A. Meyerhoff, *Newsweek* (3 March 1980), p. 11.
23. H. B. Taussig, *Scientific American* 207(2), 29 (1962).
24. G. Hardin, *Exploring New Ethics for Survival* (Viking, New York, 1972), p. 61.
25. L. Mumford, *The Transformation of Man* (Harper & Row, New York, 1956), p. 104.
26. *The World Almanac & Book of Facts 1980* (Newspaper Enterprise Assn., New York, 1979), 430.
27. *Ibid,* p. 429.
28. J. Mander, *Four Arguments for the Elimination of Television* (Morrow, New York, 1978). It is a measure of the privileged power of the media that there are few full-length attacks on the medium that devours more than half of the public's disposable time: a fine and frightening example of positive feedback in a social system.
29. J. Milton, *Areopagitica* (1644).
30. B. F. Skinner, *Beyond Freedom and Dignity* (Knopf, New York, 1972); H. Wheeler,Ed., *Beyond the Punitive Society* (W. H. Freeman, San Francisco, 1973).
31. S. Chase, *The Economy of Abundance* (Macmillan, New York, 1934), p. 310.
32. N. Rescher, *Distributive Justice: A Constructive Criticism of the Utilitarian Theory of Distribution*

(Bobbs-Merrill, New York, 1966), p. 96. (Italics in the original.)

33. I. Kristol, *Wall Street Journal* (22 November 1979), p. 22.
34. K. E. Henion II and T. C. Kinnear, Eds., *The Conserver Society* (American Marketing Society, Chicago, 1979).
35. C. L. Schultze, *The Public Use of Private Interest* (Brookings Inst., Washington, D. C., 1977), pp. 17, 73.
36. G. Hardin, *Ann. N. Y. Acad. Sci.* 300, 87 (1977).
37. For valuable criticisms of the developing manuscript I thank Karl Henion; also Mortimer Andron and other members of the Little Men's Marching and Deep Thinking Society.

Herman E. Daly

8. Postscript: Unresolved Problems and Issues for Further Research

The reader of the foregoing articles will have noted numerous disagreements among authors regarding such fundamental issues as an energy theory of value, the viability of a solar energy techology, the propriety of discounting, whether the domain of market prices should be restricted or extended, and other unresolved issues as well. The range of disagreements at the conference itself was even greater than it appears in this volume, because of the participation of members of the audience, and because of the paper presented by Barry Commoner. Commoner's paper is not included in this collection because he did not submit it for publication. Rather than try to summarize Commoner's position, which has already been widely published elsewhere, we will simply refer the reader to his other work.

As a result of having read the papers more than once, and having listened to the discussions at the AAAS meeting, and having had the opportunity to talk further with most of the authors, I have come up with a few editorial reflections on the major issues of: (1) An Energy Theory of Value; (2) The Contested Viability of a Solar Energy Technology; (3) The Temporal Dimension and the Propriety of Discounting; (4) Demand Control and Limitation of Scale. Each of these will be discussed in turn.

Energy Theory of Value

Three distinct attitudes are expressed by the preceding authors on this subject: Costanza advocates and defends an energy theory of value; Georgescu-Roegen and Arrow reject it; Hannon recognizes that single factor theories of value are unsatisfactory for explaining economic behavior, but argues that they may provide useful insights when scarcity problems arise on critical inputs--as is now the case with energy. Although by and large I believe that Georgescu-Roegen and Arrow are correct on this question, I nevertheless think that economists have something to learn from Costanza and Hannon.

Costanza's analysis exposes the excessive reliance that orthodox economists put on factor substitution. If we count labor, government, capital, and energy as primary inputs then we can have substitution of one factor for another. We can, according to neoclassical production theory, reduce the energy input while keeping output constant, simply by increasing the other factors. How far can such substitution be carried? The usual assumption is that it can be carried quite far, and indeed in Cobb-Douglas production functions substitutability is infinite. As long as the factor in question is present in some infinitesimal (but non-zero) amount, output can be maintained constant by substituting other factors. Costanza points out that substitution breaks down if factors are not independent. If it takes energy to increase labor or capital then in reality labor and capital cannot be considered simple substitutes for energy. The increased energy required to produce more labor and capital must be taken into account, and when this is done the scope for substitution of energy is much reduced. Specifically, Costanza finds that when the energy costs of producing labor in the household are taken into account, the intersectoral variations in energy use per dollar of GNP are greatly reduced. Therefore there is less scope for "uncoupling GNP from energy" than some have assumed. This point can be made independently of any position on the energy theory of value. Likewise Costanza's exposition of the thermodynamic concept of structure and gradient as <u>necessary</u> conditions for the production of valuable commodities is cogent and does not imply an energy theory of value, since the latter affirms that embodied energy is <u>both</u> a necessary <u>and</u> sufficient condition for explaining relative value. Also, those growth economists who argue that "exponential growth in technology" will free economic growth from material and energy constraints should explain how technical progress itself can be made without material and energy investments in research. Furthermore, mere knowledge means nothing to the economic system until it is embodied in physical structures. And not all matter--energy is capable of receiving the imprint of human knowledge and purpose--only low-entropy matter-energy is capable of embodying knowledge in its structure. And low entropy matter-energy is increasingly scarce. The "exponential growth of technology" argument is just "Maxwell's Demon in a business suit" as Costanza wittily puts it.

None of the above points requires an energy theory of value, and economists must not be allowed to dismiss these insights along with the energy theory of value which Costanza also defends, in my view quite unsuccessfully. Costanza has the special virtue of being more explicit and intelligible in his defense of the energy theory of value than other "energy dogmatists," especially those of the Odum school. The reader will already be familiar with Georgescu-Roegen's objections to the "energy dogma," which I find cogent, but will not repeat. Further objections are listed below.

1. To what extent is Costanza's empirical result of rough equality in sectoral energy intensities per dollar of output (as an energy theory of value would predict) really a true empirical finding, and to what extent is it the imposed result of the analytical framework of an input-output model with only one primary input, namely energy? I think it is the latter. The input-output model presupposes an energy theory of value in its formulation, and therefore the empirical outcome, while interesting for other purposes, cannot possibly constitute evidence for or against the energy theory of value. Costanza recognizes this issue and concedes that on paper one could just as well come up with a capital or labor theory of value. His only defense is to ask rhetorically, "Can anyone seriously suggest that labor creates sunlight! The reverse is obviously more accurate." Fair enough, but one can seriously suggest that sunlight does not create basic materials, and that both materials and labor (as well as space, which is limited) are necessary to capture current sunlight, and that sunlight captured for trivial purposes yields less value than that captured for vital purposes. Just as the economists' assumption of infinite substitutability of capital, labor, etc., for energy is unrealistic, the energy theorists' assumption that energy is the proper common denominator of all resource scarcity is likewise unrealistic. Sunlight, or more generally low-entropy, is a necessary condition for something to be valuable, but it is not sufficient to explain the relative value of different low-entropy structures or processes designed for satisfying wants of differing intensity and importance, and requiring differing material qualities not reducible to the common denominator of energy, and not substitutable by finite amounts of energy within finite time periods.

2. Probably the most surprising and disturbing implication of the energy theory of value as expounded by Costanza (and in this I believe he is representative of the Odum school) is its extension of market prices to the valuation of ecosystem services heretofore considered outside the domain of the market. The empirical result (or

analytical imposition) that market prices closely reflect embodied energy is taken as a sanctification of the market within the framework of the energetic dogma. Within an overall naturalistic philosophy of energetic determinism it could not be otherwise. This is a point we will develop later. Since embodied energy values are considered accurate indicators of market values, and vice versa, market energy prices might be used to evaluate natural ecosystem structures and processes according to the amounts of energy embodied in them. As Costanza puts it, "from the ecological perspective markets can be viewed as an efficient energy allocation device which humans have evolved to solve the common problem of all species--survival." And further, since embodied energy values are accurate indicators of market values where markets exist, they may be used "to determine 'market values' where markets do not exist, for example, in ecological systems."

Some writers (Hannon in the present volume) seem to think of energy accounting as a cross-check on a fallible, distorted, and short-sighted market system which is especially useful in a period of critical scarcity of energy. Costanza seems to view market prices as determined by embodied energy and hence incapable of the kind of divergence that worries Hannon. One question not addressed by Costanza is whether prices reflect energy values under purely competitive markets or only under the particular mix of monopoly, monopsony and government control prevailing in 1967, the year of the energy input-output table on which his calculations are based. It is ironic that most economists present seemed willing to accept limits on the market so as to protect ecological values, whereas the ecologist was willing to extend market values into the ecosystem because he considered both to be coherent parts of a larger natural system determined by energetic principles. This view is a logical consequence of a thorough-going energy theory of value, and Costanza has the intellectual honesty to follow it through. But to me it is just one more reason to abandon an energy theory of value. I would certainly not want to evaluate a hurricane according to the market value of its embodied energy. Nor would I feel obliged to advocate nuclear power even if it had a greater net energy yield than all alternative technologies. Value derives from the enjoyment of life, and there is more to life than energy, or even net energy or embodied energy. For example, there is material, there is time, there is purpose, there is beauty . . . . These things all affect value without bearing any determinate relation to energy.

3. Even in its own terms of calculating the "solar energy necessary directly and indirectly to produce" all commodities, the actual accounting of embodied energy is

very incomplete. It counts only solar energy entering into agriculture, forestry, and fisheries. But solar energy obviously enters all production processes by providing light and heat. The market value of the energy required to light the earth and to warm it from near absolute zero to its actual temperature is probably greater than world GNP. How this enormous joint cost could be allocated among all its joint products, both economic and ecological products, is beyond my imagination. Furthermore, those production processes requiring low temperatures and darkness must actually work against sunlight. In these cases should sunlight count as a negatively valued input? If so, what happens to the energy theory of value? Also, the conversion of fossil fuel into equivalent sunlight by a factor of 2,000 seems not only rather arbitrary but completely abstracts from the time dimension of the biogeochemical processes by which that coversion is actually carried out. But then an energy theory of value does not allow for value to be imputed to time.

I agree with the energy theorists that the sun is the source of energy upon which all life depends. But, as far as we know, even the ancient Mayas who worshipped the sun made no attempt to relate the exchange value of their artifacts to the amount of sunlight required for their production.

4. Unlike the Mayas modern energy dogmatists usually hold an underlying naturalistic philosophy of strict biophysical determinism. They would argue that ancient Mayas and modern Americans both tend to trade artifacts and services at ratios determined by relative embodied energy, whether they realize it or not. The fact that this naturalistic determinism is expressed in energy terms does not remove the basic self-contradiction of determinist philosophy. In view of the tendency of economists virtually to ignore the natural world it is undeniable that a strong naturalistic corrective is required--Costanza, Umana, Georgescu, Hannon, and Hardin all agree on this point, and even Arrow makes no attempt to defend standard economics on this score. As Costanza says, "The required perspective is an ecological or systems view which considers man to be a part of and not apart from his environment." Certainly man is a part of his environment and this needs unparsimonious repetition. But I believe it is a mistake to go on and say that man should not be considered apart from the rest of the natural world--as if everything human were reducible to physical principles wrapped up in mechanical models. This is, of course, precisely the accepted modern view of scientific materialism, and the energy theorists have swallowed it whole.

It must be recognized, however, that this philosophy of determinism does not at all square with moral concern about ecological destruction or with the advocacy of policies to avoid the worsening energy crisis. At the close of a recent speech on such dire topics, I was taken to task by a logically consistent naturalist. "All these crises you are talking about," he said, "are nothing other than the natural process of ecological sucession viewed with anthropocentric alarm. Our evolutionary sucessors have as much right to the world as we do, and in any case the system of nature, of which we are merely a part, runs on its own naturalistic laws, and we would be powerless to change it even if you had given us any good reason for doing so, which in fact you have not." Georgescu-Roegen reports that he has come up against similar attitudes, "so forcefully presented that one could have gotten the impression that the aim was the survival of the environment, not of the human species."

Here at least was a logically consistent naturalist who, unlike many of his colleagues, did not preach naturalistic determinism on MWF while urging responsible and ethical ecological policies before congressional committees on TTS. My answer, of course, was to deny the underlying philosophy of naturalistic determinism, as anyone must if he is to speak ot responsibility, or even if he is to trust his own rational processes, which he would certainly be foolish to trust if he thought they were completely determined by arational events. Reason cannot accept the view that it is itself fully explainable by and reducible to arational causes, without immediately losing all authority. Indeed, to the extent that an action or a thought can be explained physically, we tend to discredit it. For example, "he was drunk when he said that." As J. B. S. Haldane noted, if my thoughts are determined by the random mechanical collisions of atoms, no matter over how long an evolutionary time period, then why should I consider any of my thoughts as true, including the thought that my thoughts are mechanically determined? The only way determinists have found to deal with this self-contradiction is to ignore it.

The energetic dogma suffers from an even more severe case of mechanistic determinism than does neoclassical economics, and when this fact is recognized, it is no longer so surprising that the energy theory of value results in the same laissez-faire conclusion that whatever is must be right (because it could not, according to theory, be otherwise). Costanza tells us on p. 119 that "the flow of energy has <u>not</u> been the primary concern of mainstream economists," and yet he is led to conclude toward the end of the paper that, "without realizing it, the flow of energy <u>is</u> the primary concern of economics." Although it was no part of our intention, economists have been led, as if by an energetic invisible hand, to optimize the flow of energy--if one may paraphrase Adam Smith. Why then should we strive to rethink economic theory? How could there be an energy "crisis"?

Perhaps my point, and it is worth insisting on, can be made clearer by a pair of quotations from Frederick Soddy, a pioneer in energy economics to whom Costanza also refers. Soddy gave the material world its full due, yet resisted the prevalent monistic obsession of reductionistic-determinism, and stopped short of proclaiming an energy theory of value. For the benefit of over-abstracted economists Soddy wrote:

". . . life derives the whole of its physical energy or power, not from anything self-contained in living matter, and still less from an external deity, but solely from the inanimate world. It is dependent for all the necessities of its physical continuance primarily upon the principles of the steam engine. The principles and ethics of human law and convention must not run counter to those of thermodynamics." (p. 9 Cartesian Economics, Henderson's, London 1922).

Note that Soddy says human convention and ethics cannot contradict the laws of thermodynamics, which is very different from claiming that energetic principles explain human conventions of valuation. A good chess player cannot break the rules of chess, but knowing the rules does not make one a grand master. For the benefit of energy determinists Soddy commented:

"I cannot conceive of inanimate mechanism, obeying the laws of probability, by any continued series of successive steps developing the powers of choice and reproduction any more than I can envisage any increase in the complexity of an engine resulting in the production of the 'engine driver' and the power of reproducing itself." (p. 7, Ibid.)

One might question whether reproduction is not within the domain of mechanism, but the real "miracle" is that inanimate mechanism should by itself give rise to choice, to rational and moral thought, and further that the authority of thought and value should be expected to survive the "exposure" of its arational roots!

I agree with Soddy that the fundamental problem is a kind of reductionistic monism that afflicts the scientific mentality. The neoclassical economists reduce value to psychological categories of pure preference and choice, forgetting about physical laws and ending up with absurd theories of perpetual economic growth that are contrary to the physical facts of finitude and entropy. But in combatting this error it seems to me that some of the energy theorists have moved to the opposite reductionist monism of an energy theory of value and an accompanying naturalistic determinism that cuts the logical grounds from under any moral concern or political advocacy, and is rationally untenable as well. Hannon's very modest claim for the limited usefulness of an energy theory of value in periods of critical scarcity of that factor is as much as can be reasonably said in its favor. Even in the hands of such a competent theorist as Constanza the energy theory of value creates more problems than it solves.

The Disputed Technological Viability of Direct Solar Energy Collection

Georgescu-Roegen has introduced the important distinction between a "feasible recipe" and a "viable technology." A feasible recipe refers to anything that we know how to do, with no consideration given to the structure of inputs and relations to the other sectors of the economy. A viable technology refers to a feasible recipe that is sustainable or non-parasitic in its relations with the rest of the economy: one that can replace its own physical structure out of its gross product and still have a positive net product and do this without obliging some other sector to lose its viability. Although we often hear that solar technology is already here, Georgescu-Roegen disagrees:

"The truth is that only feasible recipes are here, a viable technology not. For such a technology we need a recipe that will produce the convertors of solar radiation with the aid of only the energy converted by them. All solar recipes _known at present_ are parasites on the current technologies and therefore, will cease to be applicable when their host is no longer alive." There was certainly a divergence of opinion on this point! Barry Commoner held that fossil fuel technology is parasitic and that all sensible solar technologies are either economic already or could be made so with the stroke of a pen! Georgescu compares the logic of his position to the rationale of thermodynamics: "The only reason why perpetual motions are declared impossible is that no one has ever constructed one. Similarly the present recipes do not constitute a viable technology because no one has been able to produce solar collectors without using other sources of energy."

Several questions arise. Could it not also be said that "fossil fuel collectors" (oil wells) do not constitute a viable technology because no one has been able to produce fossil fuel collectors without using other sources of energy? Surely some solar energy is used in the production of oil field equipment. And in the early days of fossil fuel collection technology, the reliance on other sources of energy must have been far greater than it is today. Pointing to the simple non-existence of a direct solar collector that actually replaces itself using only energy generated by itself is such a stringent test that, if applied to fossil fuel technology, it too would be judged nonviable! Yet it seems clear that fossil technology could be viable without the solar input, but not so clear that solar could be viable without the fossil fuel input! This may be because solar technologies have not yet had a chance to develop. Conceivably they could be viable, but less productive in the short run than fossil fuel technologies and consequently not in use now. The viability of solar energy technology would seem to be an empirical question requiring calculations of the kind that Hannon advocates. Georgescu-Roegen's rather abstract empiricism seems insufficient to test his very challenging hypothesis, and one must share Arrow's surprise at Georgescu-Roegen's pessismism on this point. One of the aims of the organizers of this conference was precisely to subject Georgescu-Roegen's hypothesis to the kind of empirical testing that Hannon has advocated. (Another aim that did not work out as the organizers had hoped was to set up a confrontation between standard economics--represented by Kenneth Arrow--and Georgescu-Roegen's critique of standard economics. According to Georgescu-Roegen, standard economics has little regard for the second law of thermodynamics and for the singular importance of natural resources as the source of the entropic flow by which the economic system lives off the environment. Arrow did not deal with this larger issue, commenting that he was really unable to find the essential point which orthodox economics was supposed to have missed in taking account of natural resources. There remains an enormous difference in perspective which did not come into direct focus.) However, in Hannon's judgment the data were not sufficiently trustworthy or complete to support a clear conclusion on policy issues, of which the viability of direct collection of solar energy is certainly the most important one. Since Georgescu-Roegen rejects the practice of discounting in long run social decisions, such as choosing an energy technology, the most relevant calculation or R (return on net investment) in Hannon's Table 2 is that with the discount rate (λ) set equal to zero. All direct solar technologies have R values (energy output divided by energy processing cost) greater than one (ranging from 1.9 for flat plate collectors, to 6.7 for photovoltaics) which would seem to indicate the technological viability of solar energy, were it not for Hannon's explicit counsel not to take the numbers very seriously.

One of Arrow's criticisms of Hannon's methodology is relevant at this point. Hannon counts as cost only the energy used in the transformation of the basic source of energy, and does not count the basic flow of energy being transformed. This seems proper for solar energy, since so much of it appears to go unused, and since transformation of today's sunlight leaves tomorrow's sunlight undiminished. But this is not the case for non-renewable fossil fuels. Thus Hannon's methodology leaves out a major advantage of solar energy. Yet even so, Hannon's solar energy R values compare favorably with many fossil fuel technologies.

The basic reason for Georgescu-Roegen's doubts about the viability of direct solar energy collection stems from the large materials requirements made necessary by the dilute and dispersed nature of sunlight. A critical determinant of viability therefore would seem to be the degree of concentration of the required materials already existing in mines. A solar energy collector made of materials available only in geologically lean deposits is clearly at a disadvantage compared to a collector made from materials already concentrated in rich ores. Over time, as the richer deposits are depleted first, it would seem that the material structure of <u>all</u> energy convertors would become harder and harder to reproduce, so that what had been viable technologies would more and more become merely feasible recipies, parasitic on other sectors. Fossil fuel technologies, in addition to replacing their material structures out of ever leaner mines, must also extract their basic direct energy input out of ever leaner mines and wells. The former is also true for solar energy, but the latter is not. The intensity of sunlight, though dilute, is at least not diminishing over time.

Are biotic convertors subject to the same long run process? Are viable organic technologies destined to become mere parasitic feasible recipes as the dispersion of material trace elements becomes so great that existing biotic systems are unable to reproduce their physical structure within the fixed solar energy budget? Will materials concentration efficiency become the major criterion of natural selection among the alternative organic technologies tried out in nature? This is a question that elicits opposing views, with many ecologists holding that natural materials cycles are 100% complete--which, if true, would contradict Georgescu-Roegen's Fourth Law of Thermodynamics. Acceptance of this Fourth Law is not sufficient to lead to the conclusion that current direct solar energy collection technology is non-viable, even though it does seem to imply the eventual, very long run non-viability of all energy convertors, including the biological. Just how long run and very long run considerations should be reflected in current choices constitutes another larger and important area of disagreement and confusion, considered briefly below.

The Temporal Dimension and the Propriety of Discounting

Three attitudes toward the comparison of present and future amounts of energy were evident: (1) ignore the issue; (2) discount future energy by some rate of productivity or time preference; (3) explicit rejection of discounting for social decisions. Costanza treated energy the same in all time periods, as required by an energy theory of value. Hannon focused on the temporal dimension and made that a critical element in the calculation of R (rate of return on net energy invested). Arrow supported Hannon's effort in this direction. Georgescu-Roegen (in discussion) explicitly rejected discounting when the decision-making entity is quasi-immortal, as is the case for society. It is rational for individuals to discount the future in decisions made by them on their own behalf because individuals may die tomorrow and will certainly die within 100 years of any decision they make. When deciding on behalf of society economists must assume quasi-immortality for the interested party and resist discounting, or what is the same thing, use a zero rate of discount.

We have time preference and productivity rationales for discounting. Georgescu-Roegen's objection seems mainly against the time preference view. Discounting by a rate of productivity increase in the interest, say, of equalizing real consumption over generations does not seem to be ruled out by the recognition of the quasi-immortality of society. But the rate of productivity increase over time is unknown. The usual assumption is that it will be positive over the long run. But is this really reasonable? Does not the second law of thermodynamics by itself imply that over the very long run the growth rate of productivity will be negative? And if one accepts Georgescu-Roegen's "Fourth Law" then the implication is even stronger.

An earlier draft of Hannon's paper contained an observation which was omitted in the final draft, but which I found both puzzling and suggestive, and will take editorial license to quote.

"λ is the physical energy discount rate by which society expresses its willingness to invest or consume energy. Although a better interpretation is not clear to me, it seems that the minimum eventual λ is related to the depreciation rate of a desired physical structure. As such it is intimately related to the second law of thermodynamics." Without being sure of just what Hannon meant, I find this observation suggestive of the following notion: to maintain desired physical structures in a steady state, physical depreciation must be offset by new production which requires depletion of nonrenewable resource stocks. This is the rate at which want-satisfying capacity is being reduced (low-entropy resources converted into high-entropy wastes) over the long run. In this very long run or ultimate sense the growth rate of the stock of desired physical structures must be negative. If λ is determined by this entropic decay then we would be discounting by a depreciation rate rather than a growth rate. That means that our discount rate should be negative, which I take to imply that future energy would be worth more than present energy. This implication can be understood by the following rationale: a Btu in the distant future would be worth more than a Btu today because it would be used to maintain a reduced stock of desired physical structures, the marginal utility of which would consequently be greater than at present, for a given population size. Also the marginal cost of a Btu of energy from depleted mines and wells is greater than the current marginal cost. So future energy will be worth more per unit than present energy, and the energy discount rate should be negative!

These are very long run considerations, and for that reason there is a tendency to dismiss them as unrealistic. Several points may be made in reply. First, the effect of depreciation is continuous and, with a large enough stock of physical structures to be maintained, the absolute volume of depreciation could outweigh new production even in the short run. Furthermore, the ultimate stock that is being depreciated is the stock of nonrenewable resources, for which there is no compensatory production rate. If the scale of the stock of desired physical structures and the technology of maintenance were such that depreciation could be replaced by relying only on the sustained yield of renewable resources, then the growth and discount rates would be zero. Second, many long run predictions are far more certain than short run predictions. For example, I can confidently predict my death within 50 years, but the uncertainty becomes greater the shorter the time period. Likewise for the meltdown of a nuclear power plant. Third, from the above considerations it is clear that it is the time horizon that determines the discount rate and not the discount rate that determines our time horizon. If a very long perspective implies the necessity of a negative discount rate, then one cannot justify a short time horizon by arguing that at current positive discount rates a dollar twenty years from now is worth only a few cents and can be neglected. This is a circular argument because the time horizon largely determines whether we consider the discount rate to be positive, zero, or negative.

Another more common sense argument against the realism of the long run view is founded on the true observation that people usually take no action and show little interest in their own descendents beyond grandchildren. Therefore it is concluded that it is the revealed public will that the future beyond two generations should carry no weight in present decisions. This, I suggest, is a false inference. It is conceivable and quite logical, in fact, that someone who cares a great deal about the welfare of future generations would not find it worthwhile to do anything in particular for his own great-great-grandchildren. The reason for this stems from the simple fact that one's great-great-grandchildren are also the great-great-grandchildren of 15 other people in the present generation, identities as yet unknown. On the assumption that each of these other 15 co-progenitors will exert as great an influence as one's self on the common descendent (either genetically or culturally), it would hardly be worthwhile to take individual action. And clan or family measures are unrealistic because potential co-progenitors are unknown. In general a given person in a given generation will have 2^n ancestors in the nth previous generation. The anonymity and exponential fanning out of ancestry with each generation obscures and dilutes responsiblity, and consequently provision for the distant future becomes <u>a public good</u> which sensible people would seek through collective social arrangements, not through individualistic maximizing behavior. Therefore the absence of individualistic action on behalf of distant descendents cannot be taken to reveal a short time horizon because of the public good nature of provision for the distant future.

A further point can be made. My great-great grandchild may also be your great-great grandchild. Therefore, to the extent that I care for my great-great grandchild I should also care for you and for every other potential co-progenitor from whom my descendent will inherit, for good or for ill, as much as he or she will inherit from me. The duty of brotherhood and mutual caring among people in the present generation does not rest on this consideration alone, by any means. Rather this is an additional dimension of brotherhood--the recognition of potential co-progenitorhood with all people of the present generation. This point is important because concern for the future is often seen as blunting or diverting our ethical concern away from the more pressing problems of present injustice. Considered logically, however, ethical concern for the distant future should strengthen rather than weaken the bonds of mutual caring within the present, because we are all potential co-progenitors of each other's descendents.

There is a further confusion in the treatment of time by neoclassical economists: there is no distinction between intertemporal allocation and intertemporal distribution, whereas in static price theory the distinction between allocation and distribution is crucial to the very notion of Pareto optimality and efficiency. Allocation refers to the apportioning of resources among alternative uses or products on the basis of a given distribution. Distribution is the apportioning of resources (or their commodity transformations) among different people. Neoclassical economists treat all intertemporal apportioning of resources as a matter of efficient allocation determined largely by the interest rate. This makes sense for an individual allocating resources over different temporal stages of his own lifetime. But different generations are necessarily different people, so between generations we have distribution, not allocation. Distribution is a matter of justice, not efficiency. It is a function of ethical values, not the interest rate. Forcing all intertemporal division of resources into the Procrustean bed of efficient allocation and not even recognizing the concept of intertemporal distribution is a major source of confusion.

I think the proper attitude is to seek sustainability directly, for ecological and ethical reasons, while accepting the consequences for scale of the economic system and population, and for the nature of the resource base (renewable) that technology must come more and more to rely on. This leads us to the discussion of demand control and limitation of scale, which follows. The neoclassical objection to direct control of scale and demand, that it bypasses the beautiful intertemporal efficiency mechanism of market interest rates and discounting, can be set aside in view of the forgoing discussion of its basic circularity, lack of recognition of the public good nature of provision for the future, and confusion of allocation with distribution in matters intertemporal.

Demand Control and Limitation of Scale

In commenting on Georgescu-Roegen's pessimism regarding the viability of direct solar energy collection, Arrow observes that, "The viability of a system of purely renewable energy would have to be investigated empirically . . . . One cannot be sure of the outcome until the calculation is done, but I am a bit surprised at Georgescu-Roegen's pessimism. It seems very likely to me that even with existing techniques a world run purely on renewable energy (a stationary state) would be feasible. No doubt it would be a considerably poorer world than we now have, but we might expect technological improvements to remedy that." Arrow goes on to concede that, "There is no way of avoiding Georgescu-Roegen's other objection to the stationary state, the inevitable degradation of matter. It can be balanced only by increasing economies in the use of matter."

As already indicated in section II, I share Arrow's surprise and doubts about Georgescu-Roegen's non-viability hypothesis for direct collection of solar energy. Likewise I believe that a stationary state would be feasible, even while acknowledging that because of the degradation of matter a steady state is not forever. For the same reason, a stationary population is not forever, yet it makes sense to demographers to use it not only as an analytical construct, but also to advocate it as a policy goal. It is precisely by accepting the discipline of a physical steady state that we will be led to focus technology on the qualitative and informational dimensions which Arrow points to as "balancing the inevitable degradation of matter."

While accepting the long run feasibility of a steady state, Arrow notes in passing that such a world would no doubt be considerably poorer than the present one, even though he expects technological improvements to remedy that. I am doubtful that technological improvements will allow us to maintain the existing population at present levels of affluence and comfort. But that does not mean that a steady-state world would be considerably poorer than today's world. As Garrett Hardin points out, there is a trade-off between population size and prosperity. For any given level of technology we can share the renewable resource yield among many or few people. This is an obvious point and I doubt that Arrow would deny it. It is customary, as in his statement, to assume the existing population size as a fixed reference. But that customary habit of thought, reinforced by all sorts of verbal blinders such as speaking always of a "shortage of supply" rather than a "longage of demand," is exactly what Hardin wants to focus our attention on. "In a world of limits," he says, "we can become wealthy only if we subject ourselves to the discipline of demand control." This is not easy, because, as biologist Hardin states, "It is hard for selection to favor organisms that are willing to curb their demands, in competition with organisms that will not." This means that moral persuasion at the individual level, though necessary, is not a sufficient condition for limiting demand and the scale of human activities to the sustainable limits of the ecosystem. Institutional collective action, "mutual coercion, mutually agreed upon" is also required.

At this point in the reasoning hysteria usually erupts and learned holders of professional Chairs of Urban Values in New York City profess their urban and anti-ecological values from the pages of the Wall Street Journal (see the quotation from Irving Kristol in Hardin's article). Professor Kristol tells us that "it is only so long as economic growth remains a credible reality that democracy will remain an actuality." Many economists agree with Kristol. Newsweek (Sept. 8, 1980, p. 40) quotes government economist Barry Bosworth on the dangers of a decline in output, "This society has never really had a consensus on anything except the notion that 'I'm doing better than I did before' . . . . That's what kept the system together." That is an extraordinary statement. If our society is really so bankrupt morally that social cohesion derives only from the perception that "I'm doing better than I did before," then its demise is not greatly to be lamented. Historically the bases for cohesion have been much deeper and stronger than that, but admittedly they are being rapidly undermined, not least by social scientists of positivist and determinist persuasion who have no place in their reductionist models for anything other than atomistic individuals mechanically knocking against each other like balls on a billards table.

I do not deny that zero growth will strain democratic institutions, but I'm quite sure it will strain collectivist institutions just as much, and will be considered "dangerous and irresponsible talk" in the Soviet Union also. I can easily imagine a Soviet professor of urban values (if they have such a category) arguing that without growth and the resulting abundance that provides the material conditions for the emergence of the new socialist man, the whole socialist state would be torn apart by the reassertion of bourgeois individualism. The weakness of Kristol's position is that he uses an argument based on questionable political premises to counter a completely independent argument based on much more solid biophysical premises. If a biophysical impossibility confronts a political "impossibility," I think we are well advised to invest our efforts in overcoming the political "impossibility" rather than trying to recreate the biophysical world on different principles. The strict economic determinism of Kristol's view, shared by his hypothetical Soviet counterpart, indicates a curious willingness to submit to economic "laws" while resisting the much more solid biophysical laws.

In any case the past association of growth with freedom in the days of the cowboy economy will not persist in the limited spaceman economy. The big enemy of freedom and civil liberties is crisis and breakdown. The attempt to force growth in the face of increasing natural, social, and moral resistences to growth will require larger, more centralized and powerful technologies and a lot more of pushing little people out of the way. Witness the opposition to nuclear power and other big technologies by people who are tired of being pushed around in the name of growth and progress. It is no accident that nuclear power is most advanced in those countries that tolerate the least public participation and individual freedom.

It is unquestionably true that distributing a fixed pie is politically more difficult than distributing a growing one. But to conclude that "growth is a crucial precondition for democracy" is to reveal both a mystical devotion to growth and a rather mean conception of democracy. It also presupposes the possibility of continual growth, which is of course the main point at issue.

While Garrett Hardin is the only one who explicitly addressed the question of demand control and limits to scale, the issue is implicit in other papers. Georgescu-Roegen attributes past economic expansion to Prometheus I (fire, wood age) and Prometheus II (coal and the steam engine, fossil fuel age). Our choice now, he says, is to wait for the uncertain Prometheus III, or to change from the present high level of industrial activity to something analogous but not identical to that of the Wood Age. This I take to imply a large reduction in scale of population and production, and consequently to imply control and limitation of scale and demand.

Even if Prometheus III were to arrive tomorrow, however, and make possible another chain reaction of growth, would we not still be wise to chain up Prometheus' tendency to provoke the chain reaction of growth up to some new limit? Would we not be better advised to take the benefits of Prometheus III in the form of a longer life for a comfortable steady state at less than the maximum possible scale? If so, scale limitation and demand control should be pursued in either case--in the former because necessary, in the latter because desirable in the interest of sustainability or permanence.

Georgescu-Roegen recognizes that the market mechanism cannot by itself limit the scale of economic activity to conform to biophysical limits. Quantitative restrictions have had to be imposed in order "to preserve forests, fish, game--or a healthy environment." To those who would argue that such restrictions on growth are inconsistent with democracy, I would mention the case of game quotas for hunters. Even that most individualistic breed, the Louisiana duck hunter, has accepted the need for quotas limiting the scale of this activity to fit the sustainable capacity of nature. This offers some empirical support to those who believe that a more generalized demand limitation is not inconsistent with democracy.

The question of how to limit scale and growth, whether by depletion quotas auctioned by the government, or by national ad valorem severance taxes, or energy taxes is an important area for further research, but was not dealt with in this session. (Interested readers are referred to H. E. Daly, ed. Economics, Ecology, Ethics, W. H. Freeman Co., San Francisco, 1980).

The growth ideology of standard economics and the consequent unwillingness seriously to consider natural and social limits to the scale of human economic activity stem from two sources: (1) the Keynesian emphasis on investment to stimulate aggregate demand in order to maintain full employment; and (2) the utilitarian philosophy on which economic theory was founded, especially Jeremy Bentham's dictum that the goal of policy should be "the greatest good for the greatest number." Let us consider each in turn.

(1) One of the founders of modern growth economics, Evesy Domar, starting from Keynesian premises, reasoned his way to a very disturbing conclusion: "In a private capitalist society where the [marginal propensity to consume] cannot be readily changed, a higher level of income and employment at any given time can be achieved only through increased investment. But investment as an employment creating instrument is a mixed blessing because of its effect [in increasing productive capacity]. The economy finds itself in a serious dilemma: if sufficient investment is not forthcoming today, unemployment will be here today. But if enough is invested today still more will be needed tomorrow . . . . So far as employment is concerned, investment is at the same time a cure for the disease and the

cause of even greater ills in the future." ("Expansion and Employment," American Economic Review, March 1947).

In other words, to solve the problem of unemployed productive capacity we must build still more productive capacity. A full employment GNP must be a growing GNP (whether it consists of bread or refilled holes in the ground). Even though this conclusion follows logically from Keynesian abstractions, Domar considered it a "serious dilemma," and so do I. It is disturbing because, like so much of economic theory, it abstracts both from the final cause of economic activity (is more GNP always worth the effort?) and from the material cause (are resources infinitely available in the aggregate?) Instead of an answer to the logical dilemma of an infinite regress, growth economics has substituted the psychological delusion of infinite progress! That which should have been an intellectual embarrassment has become the most celebrated of political axioms. Economic growth is hailed as a "crucial precondition for democracy," and growth forever has become the accepted wisdom by which today's policies are guided. But what happened to Domar's "serious dilemma"? Who has resolved it?

No one has resolved Domar's Dilemma. We are alternately impaled on one horn and then the other. If for the sake of employment we push investment and growth in the face of increasing natural and social resistances to growth, we get inflation along with pollution and rapid depletion. If we restrain aggregate demand to fight inflation we get unemployment and greater inequality in income distribution. Moreover, the tendency of increasing scarcity is to worsen income distribution. As energy and other resources become more scarce resource owners will receive larger rents, which will have to come at the expense of either capital or labor. But capital will be in ever greater demand because of the high capital intensity required for mining and refining ever leaner ores, because of the high capital intensity of most new energy technologies, and because of the tendency of technology to substitute capital for labor so as to reduce the vulnerability of the enterprise to strikes. The tendency will be for the larger share of income going to ownership (rent) to come at the expense of a smaller share to labor. The obvious solution is to capture the increasing rent and redistribute it as a kind of social dividend, or minimum income, independent of and supplementary to wage income. Rent cannot be abolished. It is a legitimate price that must be paid if we are to have an accurate accounting of the true costs of increasing scarcity. But since rent is a payment in excess of necessary supply price, its increase need not accrue to the resource owner. It could be paid to the government and redistributed without distorting market efficiency and the proper accounting of scarcity. Redistribution, either of income or of asset ownership, seems to offer the only escape from worsening income distribution and ever more frantic attempts to grow in order to provide more jobs, in order to distribute income to those who do not own resources or capital. Limitation of scale to fit within biophysical constraints requires limitation of the degree of inequality in the distribution of income. Market forces left to themselves will distribute resources more unequally as a consequence of increasing scarcity rents and capital intensities.

(2) The utilitarian goal of "the greatest good for the greatest number" has obvious pro-growth implications which were built into the philosophical foundations of economics right from the beginning. There are, however, three problems with Bentham's dictum: the definition of "good;" the logical impossibility of double maximization; and the ambiguity of number--does it refer to number alive now, or at some future time, or to the cumulative number ever to live over time?

Economists have tended to define "good" as "goods and services." This is a very reduced conception of the good, but has the advantage of being measurable. It may be worthwhile to adopt such a limited view of the good in the interest of measurability, but then one should not want to maximize it, precisely because it is such a partial element in any reasonable concept of the good. Satisficing rather than maximizing seems more appropriate: "sufficient goods and services . . . ."

The problem of double maximization has been widely recognized. The dictum contains one "greatest" too many. Rather than try to maximize two variables we must maximize one subject to some sufficiency constraint imposed on the other. I have already suggested that the sufficiency constraint be applied to goods and services. Thus we have "sufficient goods and services for the greatest number . . ." But number of what? Number of currently living people? Or cumulative number of people ever to live over time? I suggest the last. The corrected dictum thus becomes: "sufficient goods and services for the greatest number over time." This dictum implies growth only up to sufficiency. Far from implying maximum population growth the greatest cumulative number over time implies that the population alive at any one time be strictly limited to the biophysical carrying capacity. Any permanent destruction of carrying capacity required to support more people in the present implies a reduction in the population at all future time periods, and consequently a reduction in the total number of people ever to live. Those who favor zero, or even negative population growth, far from being "anti-life" are actually aiming to maximize the number of lives lived over time at a level of per capita income sufficient for each life to be lived fully and abundantly, but not extravagantly. It is the population growth boosters who are, in effect if not intention, anti-life.

This revised ethical dictum may seem excessively future-oriented. But really it is not, because the basic needs of the present (sufficiency) take precedence over all needs of the future. But the basic needs of the future are allowed to take precedence over the more trivial wants of the present. The revised rule, free of logical contradiction and emphasizing sufficiency and permanence, implies a steady-state economy rather than a growth economy. The steady state is not eternal as Georgescu-Roegen points out, but there is still an important difference between an economic system that will last less than 50 years and one that could perdure for 500,000 years. Developing the ecological understanding, technology, and social institutions appropriate for a steady-state economy should be the major task of biologists, physical scientists, economists, and philosophers during the coming decades.

187 -191

Robert Costanza

7230
7220
2260

Title p. 119

9. Reply: An Embodied Energy Theory of Value

Dr. Daly's postscript exhibits the literary excellence typical of all of his writings. I am unmoved, however, by his (and Georgescu-Roegen's earlier) objections to an embodied energy theory of value. Some of this disagreement is over definitions and assumptions (i.e., valid objections to an energy theory of value may not hold for an embodied energy theory), but I also perceive a note of cognitive dissonance arising from a clash of paradigms. Daly's objections seem to stem from a basic philosophical position that cannot be completely addressed here. I will try instead to respond briefly to each major issue, pinpointing the sources of disagreement.

As a system ecologist, my interest is in developing and using descriptive and analytical tools applicable with equal facility to ecological and economic systems. A quantitative common denominator is critical to achieving this goal. Many of Daly's objections seem to be derived from a deep-seated conviction of the uniqueness of humans: that humans are so unique that no physical theory can ever explain their choice and valuation behavior and, therefore, no common denominator is possible. This is the mainstream conception in economics, but in the other social sciences, it is rapidly losing ground to less "exemptionalist" paradigms (1, 2).

The human uniqueness component of the standard economic paradigm is in my view the last vestige of the man-centered conception of the universe that science has been chipping away at for the last 400 years. It is time to drop this last arrogant facade and get on with the business of understanding the world and our place in it.

Matter Matters

First, let me address an objection to an energy theory

of value raised earlier by Georgescu-Roegen (3) and alluded to in Daly's postscript. Georgescu argues that since "matter matters," an energy theory of value could never work. He defines matter as: "not just mass, but some positive amount of mass and some positive amount of energy structured in the definite patterns of the chemical elements and their compounds" (p. 1029). I submit that the concept of embodied energy (as distinct from calorimetric energy) incorporates just the distinction Georgescu requires. In fact, his definition of matter would do nicely as a definition of embodied energy (though I would generalize to include all organized structures, not just chemical elements and compounds). Embodied energy is the direct and indirect energy required (in combination with unstructured mass) to produce organized material structures. The energy embodied in material structures is taken as a measure of their degree of organization--the amount of low entropy they contain. Thus, Georgescu's objection is a valid criticism of an energy but not of an embodied energy theory of value.

Georgescu's formulation of a "fourth law of thermodynamics" (which says that complete recycling is impossible in a closed system) does not follow logically from the view that "matter matters," but is a pure assertion. Unlike the second law of thermodynamics, which is also an assertion, Georgescu's fourth law does not have over 100 years of attempted falsification to back it up. In fact, the closed microcosms that have existed for several years at near steady state in several ecological laboratories would seem to falsify the hypothesis, as would the continued existence of the closed biosphere. The assertion is also at odds with Prigogine's (4) recent Nobel Prize-winning work. At any rate we cannot, at present, take Georgescu's fourth law (hypothesis) as evidence for the inadequacy of an embodied energy theory of value.

Empiricism

Daly's first objection asks if my results are a "true empirical finding" or an "imposed result of the model." If a "true empirical finding" means a legitimate experimental result or proof, I would concede that they are not, but tests of this kind are notoriously difficult because of our inability to conduct controlled experiments on economic systems. The results do, however, indicate the true empirical implications of varying the critical, primary assumptions. If labor and government energy costs are accounted for in the suggested way, the empirical results are consistent with an embodied energy theory of value. It is important to note that the resulting embodied energy intensities are not identically equal and that there are important outliers. This implies that the results are not merely mathematical artifacts but, rather, have some empirical implications.

All I have really done, however, is to show that the critical test for the validity of an energy theory of value is at the level of the basic assumptions. To the extent that energy is, in reality, the only net input to our closed biosphere (not ignoring the complex array of intermediate inputs or the initial conditions), the embodied energy theory of value is valid. Does this require more than direct observation for proof?

Sufficiency

Daly and Georgescu-Roegen contend that although embodied energy (low entropy) is necessary for economic value, it is not also sufficient. This issue is largely definitional. I agree that embodied energy is not sufficient for market value, since ownership and markets are imperfect, and people fail to recognize the true value of non-marketed goods and services. On the other hand, I would argue that embodied energy is sufficient for economic value, broadly defined. Under this broader definition, anything that affects economic production directly or indirectly has an economic value (a "shadow price," in economic jargon). Air, water, marine fisheries, and poisonous mushrooms all have economic value (and always have had under this broader definition), but they may not have a market value, since markets for these things are incomplete and imperfect.

Market Imperfection

An embodied energy theory of value postulates that a perfectly functioning market would, through a complex evolutionary selection process, arrive at prices proportional to embodied energy intensity. The transactions covered by the national input-output tables are carried on in relatively well-behaved (though not perfect) markets, and the theory predicts a good (though not perfect) empirical relation between market prices and embodied energy intensities for these sectors. This is a positive statement about how the existing market works. Given that this statement is relatively accurate, we can make a normative statement about what market (shadow) prices would be in the absence of all imperfections. Daly's criticism that an embodied energy theory of value implies the impossibility of divergence between market prices and embodied energy intensities is not accurate. To the contrary, I see large bodies of transactions (i.e., in ecological systems) where markets are nonexistent or incomplete. In these systems embodied energy obviously diverges from market price. It is just these points of divergence that are most interesting, since embodied energy may be useful in correcting for imperfect ownership and other market imperfections.

In personal correspondence, Daly presented a hypothetical example that he felt "disproved" the embodied energy theory of value. The example is based on the economy of a prisoner of war camp. Goods arrive in Red Cross packages like manna from heaven with no production costs to the prisoners and the goods are exchanged with definite prices, based on the prisoners' preferences. Daly concludes that "if prices can exist without energy costs of production, even in a 'laboratory' economy, then it seems that energy cannot be a necessary and sufficient cause of prices. Something else is involved." This example illuminates some important distinctions. The point of the energy theory is that no low entropy structures (care packages) can exist without energy costs of production, and if these costs are not apparent to us (the "prisoners"), it is because the market is imperfect (the costs occur outside the prison walls), not because the energy costs do not exist.

The prisoners are in a position analogous to our position relative to environmental resources. Imperfect ownership and imperfect information about environmental production processes can lead to severe externalities affecting environmental resource use. Efficient allocation for the whole system requires that we make the "prisoners" aware of the true production costs of their "care packages." Instead of disproving the embodied energy theory of value, the prisoner example highlights how it may be useful in extending preference-based valuations to correct some glaring and increasingly important imperfections. Economics has held itself prisoner by limiting its view to market transactions for too long. As Daly (5, p. 398) has himself pointed out: "the 'biophysical foundations of economics remain ever present in the background'" and it is to these that the embodied energy theory of value addresses itself.

Incompleteness

I agree with Daly's third point that the actual accounting of embodied energy (especially in ecosystems) is still very incomplete. We disagree as to whether it can ever be made otherwise. The complete accounting would have no particular problems with production processes requiring low temperatures and darkness (as long as they are indirectly driven by sunlight, i.e., anerobic decomposition in the cold, dark depths of organic matter produced in the warm, bright light). A more complete accounting would also more fully incorporate the temporal dimension, which Daly incorrectly places outside the grasp of an embodied energy theory of value.

Mechanistic Determinism vs. Biophysical Evolution

Daly misrepresents the embodied energy theory of value as one based conceptually on mechanistic determinism and reductionist monism. As Umaña points out in his paper in this volume, the biophysical perspective seeks to go beyond the mechanical models that characterize neo-classical economics to ones based on irreversible thermodynamics and broadly defined evolution. Ever since Darwin's original exposition, there has been difficulty explaining the natural selection process since it does not rely on direct causality. The beaks on Darwin's finches were not caused in any direct way by the size of the seeds they ate, but rather, the environment exerted subtle but pervasive "pressure" to adapt. In the same spirit, I do not claim that embodied energy causes or determines market values in any direct, mechanistic sense. Instead, since energy is the only net input to our closed biosphere, the economic system adapts to this constraint and attempts (albeit imperfectly and unconsciously) to set prices to optimize the use of this ultimately scarce resource. Explicity stating the constraints imposed by irreversible thermodynamics goes a long way toward explaining surviving human conventions and valuations in the same way that knowing the higher level strategic constraints of chess (not merely the allowable moves of the pieces) is the essence of grand mastery. These strategic rules are well enough formulated to be the basis for computer programs that now play chess at the master's level.

Thus, the embodied energy theory of value is the antithesis of mechanistic determinism and is based on a notion of global interdependence, not reductionist monism. It does not prohibit free choice but depends on choice at various levels over various time frames to fuel the selection process. It avoids, however, the arrogance of granting exogenous primacy to human choices.

Although the embodied energy theory of value admittedly requires much further elaboration and testing, it may provide useful solutions to the common denominator problem. These solutions only make sense within an ecological paradigm. Just as the flat earth theory was a workable approximation until circumnavigation became possible, the human uniqueness paradigm is a workable approximation until the limits of our environmental resources are approached. This is the point where the "prisoners" begin to affect the production of "care packages." Beyond this point, a common denominator is essential to sucessful navigation.

References

1. Dunlap, R. E. 1980. Paradigmatic change in social science: from human exemptionalism to an ecological paradigm. Am. Behavioral Sci. 24:5-14.

2. Hardesty, D. L. 1977. Ecological anthropology. John Wiley, N. Y.

3. Georgescu-Roegen, N. 1979. Energy analysis and economic valuation. Southern Eco. J. 45:1023-1058.

4. Prigogine, I. 1978. Time, structure, and fluctuations. Science 201:777-785.

5. Daly, H. E. 1968. On economics as a life science. J. Pol. Econ. 76:392-406.

6. I'd like to thank L. M. Bahr, C. Neill, and J. Bartholomew for reviewing these comments and providing useful suggestions.

193-200

Nicholas Georgescu-Roegen

7220
7230
2260

Title
p. 43

10. Reply

It now appears that it is incumbent upon me to try to clarify two important points raised by Herman Daly in his editorial "Postscript" about my own position. Inevitably, I will have to refer to my previous works, most of which are cited in the article included in this volume.

The first concerns my "pessimism." According to Oxford Universal Dictionary, pessimism is "the tendency to look at the worst aspect of things." I believe that a more appropriate definition, especially for the present problem, is that a pessimist places greater weight on the factors that may bring about an unfavorable outcome than on those on which a favorable outcome may depend. The contrary inclination defines then an optimist. It is thus clear that if there are only favorable or only unfavorable factors, there can be neither optimism nor pessimism. If everything indicates that a cancerous metastasis has gone beyond any medical control, the physician who diagnoses terminal cancer certainly is not a pessimist.

But let's take the following case. A certain disease prevalent in a certain area is known to be contracted on the average by one individual out of ten thousand. If contracted, the mortality rate is one per ten thousand. A pessimist may not wish to visit that land because he may think only of the risk of death; an optimist, on the other hand, may look only at the small chance of contracting the disease and go there without any qualms. There is the scientifically neutral way, which is to compute the compound probability for a tourist to die of that disease, which is $1/10^8$. However, that is not the end for anyone who has to decide: "to go or not to go." For an optimist such a probability may be tantamount to never, to a pessimist it may seem high enough for not running the risk. Let me add that this very issue con-

concerns even decisions that a regulatory health agency may have to make.

The case that may enmesh us in a troublesome controversy is that in which it is totally impossible to arrive at a quantitative measure of the influence of each factor on which the future outcome depends. We know, for example, that the increase of carbon dioxide in the atmosphere produces a greenhouse effect which tends to heat our planet. But simultaneously the industrial activity discards various particulate matters that act as a screen for solar radiation. Which factor would outweigh the influence of the other is still inconclusive because of lack of reliable measurements. Here, no matter what factor one believes to outweigh the other, one should be a pessimist because either result will be unfavorable to our life. An optimist should in this case believe that the two factors just balance out, a highly improbable coincidence.

Finally, there is the situation in which one set of factors support a favorable outcome and another set an unfavorable one without any possibility for us to know their relative importance, perhaps not even all the factors that may influence the outcome. The typical case is that of predicting the future evolution of a unique complex, such as mankind. Such a task is beset by insuperable obstacles stemming from our finite temporal and spacial limits (Georgescu-Roegen, 1966; 1971; 1972; 1979a).

When it comes to predicting the technological future of mankind, standard economists still cling to the view epitomized by the refrain, "Come what may, we shall find a way" (Georgescu-Roegen, 1972). The only evidence invoked in support of this view is the undeniable fact that ever since the dawn of history mankind has been able to achieve technical progress, let alone maintaining its industrial efficiency at all times. This argument is a clear-cut manifestation of optimism. There are several reasons for this verdict. First, the meaning of the refrain is that, come what may, we shall find a way to continue to grow even exponentially, not only locally but globally as well. If jet planes can no longer be flown because fossil fuels have been all exhausted, we shall use "flying saucers" of some sort propelled by another kind of energy. Certainly, the refrain does not say simply that mankind will find a way for surviving regardless of any ecological adversity. For if this were the case, I should register as an optimist, too. Years ago I ventured the thought that mankind will probably do better than the dinosaurs, which survived for one hundred and twenty million years (Georgescu-Roegen, 1972). This thought is based not only on the fantastic potentiality of the human nature as evidenced by its past technical evolution, but also upon the plasticity of virtually all terrestrial species. The only apprehension I had about this general view concerns the possibility that reason may not guide our actions (Georgescu-Roegen, 1970, 1981).

The essential difference between my position and the Panglossian one defended by standard economists and even many natural scientists is that I insist that mankind's past evolution is not the only factor on which to base our view of the future. There also are other elements of the picture: the Entropy Law, extended so as to include the entropic degradation of any specific material element, coupled with the finitude of our accessible environment. It is hard to see why one should refuse to recognize the role of these factors in shaping mankind's future. For it is the implications of these material constraints that bring to the forefront facts of decisive importance for our understanding of mankind's exosomatic existence. First, fossil fuels represent a unique bonanza for our industrial activity. Second, as I explained in my paper, any new technology must be based on a Promethean conversion, a far from simple condition. Therefore, if I am tagged as a pessimist because I denounced an optimism based on a one-sided view of a complex problem and on a gratuitous extrapolation of an evolutionary phase, I plead guilty and with pride. For to merely cry out, "Enough of Pessimism," will not change the march of events, nor would it vindicate the Panglossian logic.

Apparently, standard economists would also charge me with pessimism because I disagree with their absolute faith in the power of the market mechanism. But such a charge can be made only rhetorically. Elementary economic considerations prove beyond doubt that the market mechanism cannot possibly lead to any tolerable allocation of resources among successive generations, let alone steer mankind away from ecological catastrophies (1). First, market prices are historically parochial, influenced by income and resource distribution, fiscal systems, technology and tastes. Second, the market behavior of humans is myopic in that they discount heavily future enjoyments, with the result that interest

---

1. In his Richard E. Ely Lecture, Robert M. Solow (1974) displayed all his well-known talent for argumentation to defend the position that the market mechanism does allocate natural resources in an economically rational way. Highly symptomatic of the present crisis of the economic profession, in his Presidential Address, Solow (1980) came around to admitting that the market cannot allocate resources among successive generations.

rates (the market price of present over future incomes) encourage extravagant depletion of resources. Third, natural resources in situ do not have a cost of production; nor can they have a proper demand price since future generations are not present to bid, too (Georgescu-Roegen, 1966, 1971, 1972, 1976c). We also know that for the protection of the air, lakes and rivers, forests, game, and fisheries, only quantitative measures, not the market, could do the job. If the whales or the cheetahs, for example, are now menaced by extinction, it is just because the corresponding market prices encourage hunting them without restriction. How unfortunate can be the directions dictated by the market mechanism has been a point hammered by William H. Miernyk (1978). It was the market prices of crude oil for a great number of years after World War II that were responsible for the Appalachian poverty of the 1960's, for the complete lack of concern in technology for energy saving, and for the neglect of developing better techniques for the use of coal.

But since I have been so often misread, probably because most readers were disturbed by my critique of the conventional faith, I wish to remind the reader this time that in my opinion the market is the best computer for establishing an acceptable allocation of resources and distribution of income, but only for the economy of a single generation. But to ensure that the fate of future generations is not completely disregarded by the current economic decisions, a universal control must be set up for the trade between the economic process and the environment. Quantitative restrictions must be set up for the import of terrestrial resources into the economic process, qualitative restrictions for the export. Between these two controls, the market must be allowed to operate freely but without impinging against basic welfare norms (Georgescu-Roegen, 1976c).

I may now turn to the second point for which I am tagged as pessimist, this time even by someone who, like Herman Daly, is as concerned with the present ecological bind as I am. Before going again over my logic, I wish to recall that I first argued in rather broad strokes that a solar technology based on the feasible recipes known-at-present cannot be viable (Georgescu-Roegen, 1978, 1978a); later, I analyzed the same idea with sufficient details (Georgescu-Roegen, 1978c, 1979a).

A crucial point of my argument is the difference between two kinds of processes: the feasible recipe and the viable technology, as these concepts were defined again in the paper of this volume. Both are defined in terms of matter-energy, not in terms of money-cost versus benefit-cost. In this connection, I should recall that for every viable technology there always is a price constellation that makes it economically workable as well; but, surprising though it may seem, a nonviable technology might become economically workable for some price constellation. These statements are mathematically demonstrable (Georgescu-Roegen, 1979a), not sweeping assertions such as those of Barry Commoner which Daly opposes to my own argument.

To elucidate the concept of feasible recipe, I mentioned that putting a man on the moon is now feasible, but controlling the fusion reaction is not. For this occasion, I should like to emphasize one point implicit in my paper. The technology based on "indirect" use of solar radiation is certainly feasible. The Wood Age, as I called it, proves the point. Actually, that technology is the basis of all life on the earth. Any other technology constitutes a superstructure. If the conversion of solar radiation developed by evolution on our planet would not exist, no other technology could exist either. Therefore, the operation of an oil well certainly depends on that energy, if it were not for other things than the men working on it. To point this fact out, as Daly does, does not refute the fact that the current fossil fuels technology is viable in the sense in which I defined it. That definition refers to a technology that is superimposed on the indispensable basis of that conversion of solar energy by "natural" channels. Any other way to conceive "viability" would be senseless. And to ignore that the current technology based on fossil fuels is a Promethean technology is to forget the unique revolution caused by the invention of the steam engine. Of course, even a viable technology, although self-sustaining, cannot continue to exist if its "fuel" is no longer forthcoming.

My proof of the fact that a solar technology is not viable at present uses a simple flow analysis of the kind on which my earlier Tables 1 and 2 are based. For the limited purpose at hand the schematic flow matrix of Table 3 would do. That matrix represents the structure of a solar energy technology (or of any other technology, for that matter). Process P_1 converts solar energy into controlled energy, CSE, with the aid of some "collectors," CL, and some other capital equipment, K. Process P_2 produces collectors with the aid of the energy controlled by P_1 and also some capital equipment; P_3 uses CSE to produce capital equipment for all purposes; and P_4 supports all other activities of production and consumption with the necessary CSE and capital equipment as well.

The conditions for viability, (3) in my main text, are

Table 3. The Reduced Flow Matrix of a Technology

Elements	(P_1)	(P_2)	(P_3)	(P_4)
CSE	x_{11}	$-x_{12}$	$-x_{13}$	$-x_{14}$
CL	$-x_{21}$	x_{22}	*	*
K	$-x_{31}$	$-x_{32}$	x_{33}	$-x_{34}$

$$x_{11} = x_{12} + x_{13} + x_{14},$$
$$x_{22} = x_{21}, \tag{3a}$$
$$x_{33} = x_{31} + x_{32} + x_{34}.$$

Obviously, if $x_{11} < x_{12} + x_{13} + x_{14}$, the technology based on the energy conversion of P_1 is not viable. But even if $x_{11} < x_{12}$, the technology is not viable. The evidence that the last inequality is at present the case is the following fact: In spite of the unusually large funds spent by public and private agencies bent on finding a solution to the unavoidable exhaustion of fossil fuels, it has not been possible to set up even a pilot plant in which solar collectors of one kind or another would be produced only with the energy collected by that conversion.

To argue that a process is impossible because no one has been able to set it up effectively is a common procedure in thermodynamics. All basic thermodynamic laws proclaim impossibilities that have never been disproved. Purists, of course, decry this logic. Yet in the ultimate analysis virtually all basic laws of physics are based on the same kind of negative evidence. Every time one has touched a hot stove, it was one's hand that was burned, not the stove (which is the quintessence of the Entropy Law). Similarly, we do not know of any case in which an object released from its hold flew up instead of falling toward the ground.

I refused to engage the distant future, one way or the other. That is, I do not want to argue, for example, that we may not discover a means to screen out gravitation some day, nor that we shall never find out how to produce collectors with only the energy collected by them. What I have endeavored to expose is the rather prevalent tendency to reason that just because we cannot deny such eventualities in the future, we must not deny their imminent realization. Only harm may result if one were to insist, for example, that the screening of gravitation is around the corner. People may thus be induced to build houses without elevators and without stairs. Similarly, if one continues to proclaim that the technology based on direct harnessing of solar energy is here, the world will grow deaf to the calls for conservation by renouncing all extravagant uses of energy until a new type of conversion is thought up. Let me also note that to point out that only feasible solar recipes are here, but that a viable solar technology is not, cannot be interpreted as a pessimist opinion. The statement is factually true.

Another point that I cannot overemphasize is that my argument is not based on the fact that solar radiation is extremely weak and, hence, its harnessing needs a large material scaffold. The weak intensity explains only one difficulty, it does not by itself prove that solar technology is not viable now. As I observed elsewhere, fission also requires a relatively large material scaffold (to contain, this time, a dangerously strong energy); yet it can very likely support a viable technology (Georgescu-Roegen, 1978c).

One can approach, however, the problem from a different angle, as some have already done by trying to see whether book data may disprove that $x_{11} < x_{12}$. Very recently, Malcolm Slesser (1981), has undertaken such an analysis. Adopting my Promethean condition for viability, Slesser came to the conclusion that even book data do not clearly disprove my position. However, even if the calculations on paper had shown that the $x_{11} > x_{12}$, the result had to be tested experimentally. That the necessary pilot plant may cost too much is not a valid defense. The cost of putting a man on the moon was immense, but that did not prevent the undertaking. No pure experiment has to be restricted by economic profitability. Of course, the complete proof of viability must include the verification of all conditions (3a).

To represent the way collectors are now produced and used, we should observe that their use is almost exclusively for heating. Their industrial uses are truly marginal instances, situations in which the use of other sources would be prohibited by technology (in spacecraft) or by cost (isolated posts on remote mountains). The present situation is represented by Table 4. All processes now use only nonsolar energy, mainly that obtained from fossil fuels, FE. A new process, P_0, must be added to represent the production of FE. In order to isolate the issues, we may assume that P_4 receives from the production complex only heat.

Table 4. The Present Technical Relations of Solar Collectors with the Associated Processes.

Elements	(P_1)	(P_0)	(P_2)	(P_3)	(P_4)
CSE	x_{11}	*	*	*	$-x_{11}$
CL	$-x_{22}$	*	x_{22}	*	*
K	$-x_{31}$	$-x_{30}$	$-x_{32}$	x_{33}	*
FE	$-x_{01}$	$+x_{00}$	$-x_{02}$	$-x_{03}$	*

We have

$$x_{00} = x_{01} + x_{02} + x_{03}. \tag{17}$$

It further stands to reason that x_{02} must be of the same order of magnitude as x_{12} of Table 3. Then, from (17) and the fact that $x_{11} < x_{12}$ it follows that x_{00} must be by far greater than x_{11}. In other words, the recipe for harnessing solar radiation is a "parasite" of the technology based on other kinds of energy. The fact that collectors are nevertheless produced and sold profitably should not surprise us or be taken as a proof of the viability of solar technology. Remember my proof that price workability does not necessarily imply technical viability.

References

Nicholas Georgescu-Roegen (c), "Economics and Mankind's Ecological Problem," The Limits to Growth, U.S. Economic Growth from 1976 to 1986: Prospects, Problems, and Patterns, Volume 7, Washington, D.C.: U.S. Government Printing Office, 1976, pp. 62-91.

________, "Solar Shading," International Herald Tribune, 12 May 1978, p. 4.

________, "The Crisis of Natural Resources," Challenge, 24 (March/April, 1981): 50-56.

William H. Miernyk, Frank Giarratani, and Charles F. Socher, Regional Impacts of Rising Energy Prices, Cambridge, Mass.: Ballinger, 1978.

Malcom Slesser, "Can Solar Energy Replace Fossil-Fissile Energy Sources?" Solar Energy, 25 (1981): 425-28.

Robert M. Solow, "On Theories of Unemployment," Am. Econ. Rev., 70 (March 1980): 1-11.